THE BIRTH OF A NEW PHYSICS

Also by I. Bernard Cohen

Benjamin Franklin's Experiments (1941)
Roemer and the First Determination of the Velocity of Light (1942)
Science, Servant of Man (1948)
Some Early Tools of American Science (1950, 1967)
Ethan Allen Hitchcock of Vermont: Soldier, Humanitarian, and Scholar (1951)
General Education in Science (with Fletcher G. Watson) (1952)
Benjamin Franklin: His Contribution to the American Tradition (1953)
Isaac Newton's Papers & Letters on Natural Philosophy (1958, 1978)
A Treasury of Scientific Prose (with Howard Mumford Jones) (1963, 1977)
Introduction to Newton's 'Principia' (1971)
Isaac Newton's *Principia,* with Variant Readings (with Alexandre Koyré & Anne Whitman) (1972)
Isaac Newton's Theory of the Moon's Motion (1975)
Benjamin Franklin: Scientist and Statesman (1975)
The Newtonian Revolution (1980)
An Album of Science: From Leonardo to Lavoisier, 1450–1800 (1980)
Revolution in Science (1985)

THE BIRTH OF A NEW PHYSICS

Revised and Updated

I. BERNARD COHEN

W·W·Norton & Company · New York · London

Published simultaneously in Canada by Penguin Books Canada Ltd, 2801 John Street, Markham, Ontario L3R 1B4.
Printed in the United States of America.

The text of this book is composed in 10/12 Baskerville, with display type set in Horizon Light. Composition and manufacturing by The Haddon Craftsmen, Inc.
Book design by Nancy Dale Muldoon.

Library of Congress Cataloging in Publication Data

Cohen, I. Bernard, 1914-
The birth of a new physics.

Bibliography: p.
Includes index.
1. Mechanics. 2. Mechanics, Celestial. I. Title.
QC122.C6 1985 530'.09 84-25582

ISBN 0-393-01994-2

ISBN 0-393-30045-5 PBK

W. W. Norton & Company, Inc., 500 Fifth Avenue, New York, N. Y. 10110
W. W. Norton & Company Ltd., 37 Great Russell Street, London WC1B 3NU

6 7 8 9 0

To
Stillman Drake
Paolo Galuzzi
Richard S. Westfall and
Eric Aiton
who have illuminated
the thought of
Galileo, Newton,
Kepler, and
Leibniz

Contents

Preface

The Birth of a New Physics has been written for the general reader, for students in high schools or colleges (studying science, philosophy, or history), for historians and philosophers, and for anyone who may wish to understand the dynamic, adventurous quality of science. I hope that scientists themselves may also find pleasure and profit in learning about the stages that led to the climax of the Scientific Revolution, the production of Newtonian mechanics and celestial mechanics.

The purpose of this book is not primarily to present a popular history of science, nor even to display for the general reader some of the recent results of research in the history of science. Rather, the intention is to explore one aspect of that great Scientific Revolution that occurred during the sixteenth and seventeenth centuries, to clarify certain fundamental aspects of the nature and development of modern science. One important theme is the effect of the closely knit structure of the physical sciences on the formation of a science of motion. Since the seventeenth century, again and again we have seen that a major modification in any one part of the physical sciences must eventually produce changes throughout; another consequence is the general impossibility of testing or proving a scientific statement in isolation or fully by itself, each test being rather a verification of the particular proposition under discussion plus the whole system of physical science.

The chief, and perhaps unique, quality of modern science is its dynamic aspect, the way in which changes constantly occur. Unfortunately, the needs of logical presentation in elementary textbooks and general works on science prevent the student and reader from gaining a true idea of this particular dynamic property. Hence another of the major aims of this book is to try to indicate the penetrating force and deep effect that a single idea may have in altering the whole structure of science.

Because this book is not a history of science, but rather a historical essay on a major episode in the development of science, it does not deal fully with every aspect of the rise of modern dynamics or astronomy. For example, Tycho Brahe's reform of observational astronomy is mentioned only in passing, as is Kepler's concept of motion and the causes of motion. A topic not treated at all is the system of Cartesian thought, including the concept of a vortex-based cosmological system. In many ways, Cartesian science represents the most revolutionary part of the new science of the seventeenth century. Other major figures whose work would have to be included in a full history are Christiaan Huygens and Robert Hooke.

I should like to acknowledge my intellectual debt to Alexandre Koyré of the Ecole Pratique des Hautes Etudes (Paris) and the Institute for Advanced Study (Princeton), our master in the scholarly art of historical conceptual analysis. Majorie Hope Nicolson (Columbia University) has made us aware of the vast intellectual significance of the "new astronomy" and particularly Galileo's telescopic discoveries. For more than a decade, to my great joy and profit, I was able to discuss many of the problems of medieval science with Marshall Clagett (University of Wisconsin; the Institute for Advanced Study), and more recently with John E. Murdoch (Harvard University) and Edward Grant (Indiana University). For almost four decades I have profited from the criticisms of Edward Rosen (City University of New York) along with his scholarly contributions. More recently, I have gained new insight into Copernican science from Noel Swerdlow (University of Chicago). I have learned much about the history and early use of the telescope from Albert Van Helden (Rice University). I have a special obligation to Stillman Drake, who over the years has been more than ordinarily generous in permitting me to see his unpublished Galilean studies and in answering questions, and who has given the typescript of this book a critical reading, first in the original edition twenty-five years ago and now once again in its revision.

The first edition of *The Birth of a New Physics,* dedicated to my daughter Dr. Frances B. Cohen, was written for the Science Study Series, part of a fresh approach to the teaching, study, and understanding of physics created by the Physical Science Study Committee, headed by Jerrold Zacharias and the late Francis L. Friedman of M.I.T. The preparation of that edition was facilitated in every imaginable way by the staff of the P.S.S.C. (notably Bruce Kingsbury); in particular I found in John H. Durston a sympathetic editor who helped me to reduce my labor to manageable proportions. I continue to be especially pleased that the photographs reproduced as plates VI and VII were specially made for this book by my old teacher and quondam student Berenice Abbott, one of America's great photographers.

The first edition has been printed and reprinted many times and has appeared in translation in Danish, Finnish, French, German, Hebrew, Italian, Japanese, Polish, Spanish, Swedish, and Turkish. The most recent of these versions, in Italian, is considerably revised and emended (including some corrections brought to my attention by Edward Rosen). Now, after an interval of some twenty-five years, the book has been updated to take account of developments and discoveries in the history of science, primarily with respect to Galileo, but also Newton. Many of the emendations and new materials have been inserted into the text, but others would have produced serious imbalances and would have destroyed the narrative pace of the original. Accordingly, the latter have been incorporated into a series of numbered supplements, referred to in the text, which amplify certain crucial issues of scholarship and understanding and which are essential to a balanced judgment concerning some of the most significant episodes in the coming-into-being of modern physical science.

Apart from the supplements, the most notable difference between the first edition and the present one is in the treatment of Galileo. During the interval between editions, we have learned (thanks initially to Thomas B. Settle's bold reproduction of one of Galileo's most famous experiments) that the experiments described by Galileo actually can give the results he claimed. Hence there has been a considerable shift of scholarly opinion. No longer is it believed that Galileo tended to describe only "thought experiments," which he either did not ever perform or could not have performed in the way he described. Rather, we have come to see Galileo as a master of the experimental art. Secondly, thanks in the greatest measure to the scholarly efforts of Stillman Drake, we have learned of the crucial importance of experiments in Galileo's formulation and testing (and even his discovery) of basic ideas on principles of motion.

I am very happy that this new edition is being published by W. W. Norton & Company. I am grateful to Edwin Barber, a vice-president, for his interest in my work. It is good to know that the world of book-making and book-selling still has a place for a real "old-time" publisher who likes books and authors.

I. Bernard Cohen

Harvard University
Cambridge, Mass.
18 September 1984

THE BIRTH OF A NEW PHYSICS

CHAPTER

1

The Physics of a Moving Earth

Odd as it may seem, most people's views about motion are part of a system of physics that was proposed more than 2000 years ago and was experimentally shown to be inadequate at least 1400 years ago. It is a fact that presumably well-educated men and women tend even today to think about the physical world as if the earth were at rest, rather than in motion. By this I do not mean that such people "really" believe the earth is at rest; if questioned, they will reply that of course they "know" that the earth rotates once a day about its axis and at the same time moves in a great yearly orbit around the sun. Yet when it comes to explaining certain common physical events, these same people are not able to tell you how it is that these everyday phenomena can happen, as we see they do, on a moving earth. In particular, these misunderstandings of physics tend to center on the problem of falling objects, on the general concept of motion. Thus we may see exemplified the old precept, "To be ignorant of motion is to be ignorant of nature."

WHERE WILL IT FALL?

In the inability to deal with questions of motion in relation to a moving earth, the average person is in the same position as some of the greatest scientists of the past, which may be a source of considerable comfort. The major difference is, however, that for the scientist of the past the inability to resolve these questions was a sign of the times, whereas for us moderns such inability is, alas, a badge of ignorance. Characteristic of these problems is a print of the seventeenth century (Plate 1) showing a cannon pointing up in the air. Observe the question that is asked, "*Retombera-t-il?*" (Will it fall back down again?) If the earth is at rest, there is no doubt that the cannon ball fired straight up in the air should eventually come straight down again into the cannon. But will it on a moving earth? And if it will, why? The plate actually illustrates an even more complex problem of motion. Here we need only note that the nature of the path of a body or projectile hurled straight upward or dropped straight downward was very early seen to be one of the intellectual hurdles in accepting the concept that the earth moves.

Suppose the earth is in motion. Then, an arrow shot up into the air must move along with the earth while it ascends and later descends; otherwise, it would strike the earth far from the archer. A ready traditional answer is that the air must move along with the earth and hence the ascending and descending arrow is carried along. But the opponents had a ready reply: Even if the air could be supposed to move—a difficult supposition since there is no apparent cause for the air to move with the earth—would not the air move very much more slowly than the earth, since it is so very different in substance and in quality? Hence, in any case, would not the arrow be left behind? And what of the high winds that a man in a tower should feel if the earth was rushing through space?

In order to see these problems in sharper relief, we can for a moment ignore the earth itself. After all, the average man and woman may very well reply: I may not be able to explain *how* a ball dropped from a tower will strike the ground at the foot of the tower even though the earth is moving. But I *do know* that a dropped ball descends vertically, and I *do know* that the earth is in motion. So there *must* be some explanation, even if I am not aware of what it is.

Let us, then, deal with another situation altogether. Simply assume that we are able to construct some kind of vehicle which will move very quickly—so quickly indeed that its speed will be approximately 20 miles per second. An experimenter stands at the end of this vehicle, on an observation platform of the last car if it happens to be a train. While the train is rushing ahead at a speed of 20 miles a second, he takes an iron ball weighing about a pound from his pocket and throws it vertically into the air to a height of 16 feet. The ascent takes about one second, and it takes another second for the ball to come down. How far has the man at the end of the train moved? Since his speed is 20 miles per second, he will have traveled 40 miles from the spot where he threw the ball into the air.

We are in a position somewhat like the man who drew the picture of a cannon firing a ball up into the air. We ask: Where will it fall? Will the ball come down to strike the track at or very near the place from which it was thrown? Or, will the ball somehow or other manage to come down so near the hands of the man who threw it that he will be able to catch it, even though his train is moving at a speed of 20 miles per second? If you reply that the ball will strike the track some miles behind the train, then you clearly do not understand the physics of the earth in motion. But, if you believe that the man at the back of the train will catch the ball, you will then have to face the question: What "force" makes the ball move forward with a speed of 20 miles a second even though the man throwing the ball gave it an upward force and not a force along the track? (Those who may be concerned about the possibilities of air friction can imagine the experiment to be conducted inside a sealed car of the train.)

The belief that a ball thrown straight upward from the moving train will continue to move along a straight line straight up and straight down, so as to strike the track at a point far behind the train, is closely related to another belief about moving objects. Both are part of the system of physics of about 2000 years ago. Let us examine this second problem for a moment, because it happens that the same people who do not understand how objects can appear to fall vertically downward on a moving earth are also not entirely sure what happens when objects of different weight fall. Everyone is aware, of course, that the falling of a body in air depends upon its shape. This can be easily demonstrated if you make a parachute of a handkerchief, knotting the four corners of the handkerchief to four pieces of string and then tying all four pieces of string together to a small weight. Roll this parachute into a ball and throw it up into the air and you will observe that it will float gently downward. But now make it into a ball again, take a piece of silk thread and tie it around the handkerchief and weight so that the handkerchief cannot open in the air, and, as you will observe, the same object will now plummet to earth. Bodies of the same weight but of different shape fall with different speeds. But what of objects of the same shape but of different weight? Suppose you were to go to the top of a high tower, or to the third story of a house, and that you were to drop from that height two objects of identical shape, spherical balls, one weighing 10 pounds and the other 1 pound. Which would strike the ground first? And how much sooner would it strike? If the relation between the two weights, in this case a factor of ten to one, makes a difference, would the same difference in time of fall be observed if the weights were respectively 10 pounds and 100 pounds? And what if they were 1 milligram and 10 milligrams?

ALTERNATIVE ANSWERS

The usual progression of knowledge of physics goes something like this: First, there is a belief that if 1- and 10-pound balls are dropped simultaneously, the 10-pound ball will strike the ground first, and that the 1-pound ball will take ten times as long to reach the ground as the 10-pound ball. Then follows a stage of greater sophistication, in which the student presumably has learned from an elementary textbook that the previous conclusion is unwarranted, that the "true" answer is that they will both strike at the same time no matter what their respective weights. The first answer may be called "Aristotelian," because it accords with the principles that the Greek philosopher Aristotle formulated in physics about 350 years before the beginning of the Christian era. The second exemplifies the "elementary textbook" view, because it is to be found in many such books. Sometimes it is even said that this second view was "proved" in the seventeenth century by the Italian scientist Galileo Galilei. A typical version of this story is that Galileo "caused balls of different sizes and materials to be dropped at the same instant from the top of the Leaning Tower of Pisa. They [his friends and associates] saw the balls start together and fall together, and heard them strike the ground together. Some were convinced; others returned to their rooms to consult the books of Aristotle, discussing the evidence."

Both the Aristotelian and the "elementary textbook" views are wrong, as has been known by experiment for at least 1400 years. Let us go back to the sixth century when Joannes Philoponus (or John the Grammarian), a Byzantine scholar, was studying this question. Philoponus argued that experience contradicts the commonly held views of falling. Adopting what we would call a rather "modern" attitude, he said that an argument based on "actual observation" is much more effective than "any sort of verbal argument." Here is his argument based on experiment:

> For if you let fall from the same height two weights of which one is many times as heavy as the other, you will see that the ratio of the times required for the motion does not depend on the ratio of the weights, but that the difference in time is a very small one. And so, if the difference in the weights is not considerable, that is, if one is, let us say, double the other, there will be no difference, or else an imperceptible difference, in time, though the difference in weight is by no means negligible, with one body weighing twice as much as the other.

In this statement, we find experimental evidence that the "Aristotelian" view is wrong because objects differing greatly in weight, or those that differ in weight by a factor of two, will strike the ground at almost the same time. But observe that Philoponus also suggests that the "elementary textbook" view may be incorrect, because he has found that bodies of different weight may fall from the same height in slightly different times. Such differences may be so small as to be "imperceptible." One millennium later the Flemish engineer, physicist, and mathematician Simon Stevin performed a similar experiment. His account reads:

> The experience against Aristotle is the following: Let us take (as the very learned Mr. Jan Cornets de Groot, most industrious investigator of the secrets of Nature, and myself have done) two spheres of lead, the one ten times larger and heavier than the other, and drop them together from a height of 30 feet onto a board or something on which they give a perceptible sound. Then it will be found that the lighter

> will not be ten times longer on its way than the heavier, but that they fall together onto the board so simultaneously that their two sounds seem to be one and the same rap.

Stevin was obviously more interested in proving Aristotle wrong than in trying to discern whether there was a very slight difference, which would have been somewhat accentuated had he dropped the weights from a greater height. His report is, therefore, not quite so accurate as the one Philoponus gave at the end of the sixth century. He did not allow for a small, but perhaps often "imperceptible," difference in time.

Galileo, who performed this particular experiment with greater care than Stevin, reported it in final form:

> But I . . . who have made the test can assure you that a cannon ball weighing one or two hundred pounds, or even more, will not reach the ground by as much as a span ahead of a musket ball weighing only half an ounce, provided both are dropped from a height of 200 braccia . . . the larger outstrips the smaller by two inches, that is, when the larger has reached the ground, the other is short of it by two inches.

THE NEED FOR A NEW PHYSICS

What, you may still wonder, has the relative speed of light and heavy falling objects to do with either a world system in which the earth is in motion or the earlier systems in which the earth was at rest? The answer lies in the fact that the old system of physics associated with the name of Aristotle was a complete system of science developed for a universe at the center of which the earth is at rest; hence, to overthrow that system by putting the earth in motion required a new physics. Clearly, if it could be shown that the old physics was inadequate, or even that it led to wrong conclusions, one would have a very powerful argument for rejecting the old system of the universe. Conversely, to make people accept a new system, it would be necessary to provide a new physics for it.

I assume, of course, that you, the reader of this book, accept the "modern" point of view, which holds that the sun is at rest and that the planets move around it. For the moment let us not inquire what we mean by the statement "The sun is at rest," or how we might prove it, but simply concentrate on the fact that the earth is in motion. How fast does it move? The earth rotates upon its axis once in every 24 hours. At the equator the circumference of the earth is approximately 24,000 miles, and so the speed of rotation of any observer at the earth's equator is 1000 miles per hour. This is a linear speed of about 1500 feet per second. Conceive the following experiment. A rock is thrown straight up into the air. The time in which it rises is, let us say, two seconds, while a similar time is required for its descent. During four seconds the rotation of the earth will shift the place from which the object was thrown through a distance of some 6000 feet, a little over a mile. But the rock does not strike the earth one mile away; it lands very near the point from which it was thrown. We ask: How is this possible? How can the earth be twirling around at this tremendous speed of 1000 miles per hour and yet we not hear the wind whistling as the earth leaves the air behind it? Or, to take one of the other classical objections to the idea of a moving earth, consider a bird perched on the limb of a tree. The bird sees a worm on the ground and lets go of the tree. In the meanwhile the earth goes whirling by at this enormous rate, and the bird, though flapping its wings as hard as it can, will never achieve sufficient speed to grab the worm—unless the worm is located to the west. But it is a fact of observation that birds do fly from trees to the earth and eat worms that lie to the east as well as to the west. Unless you can see your way clearly through these problems without a moment's thought, you do not really live modern physics to its fullest, and for you the statement that the earth rotates upon its axis once in 24 hours cannot actually have its full physical meaning.

If the daily rotation presents a serious problem, think of the annual motion of the earth in its orbit. It is relatively simple to compute the speed with which the earth moves in its orbit around the sun. There are 60 seconds in a minute and 60 minutes in an hour, or 3600 seconds in an hour. Multiply this number by 24 to get 86,400 seconds in a day. Multiply this by 365¼ days, and the result is somewhat more than 30 million seconds in a year. To find the speed at which the earth moves around the sun, we have to compute the size of the earth's orbit and divide it by the time it takes for the earth to move through the orbit. This path is roughly a circle with a radius of about 93 million miles, and a circumference of about 580,000,000 miles (the circumference of the circle is equal to the radius multiplied by 2π). This is equivalent to saying that the earth moves through about 3,000,000,000,000 feet in every year. The speed of the earth is thus

$$\frac{3{,}000{,}000{,}000{,}000 \text{ feet}}{30{,}000{,}000 \text{ seconds}} = 100{,}000 \text{ ft/sec}$$

Each of the questions raised about the rotating earth can be raised again in magnified form with regard to an earth moving in an orbit. This speed of 100,000 feet per second, or about 19 miles per second, shows us the great difficulty encountered at the beginning of the chapter. Let us ask this question: Is it possible for us to move at a speed of 19 miles per second and not be aware of it? Suppose we dropped an object from a height of 16 feet; it would take about one second to strike the ground. According to our calculation, while this object was falling the earth should have been rushing away underneath, and the object would strike the ground some 19 miles from the point where it was dropped! And as for the birds on the trees, if a bird hanging on a limb for dear life were to let go for an instant, would it not be lost out in space forever? Yet the fact is that birds are not lost in space but continue to inhabit the earth and to fly about it singing gaily.

These examples show us how difficult it really is to face the consequences of an earth in motion. It is plain that our ordinary ideas are inadequate to explain the observed facts of daily experience on an earth that is either rotating or moving in its orbit. There should be no doubt, therefore, that the shift from the concept of a stationary earth to a moving earth necessarily involved the birth of a new physics.

CHAPTER

2

The Old Physics

The old physics is sometimes known as the physics of common sense, because it is the physics that most people believe in and act upon intuitively. It is the kind of physics that seems to appeal to anyone who uses his native intelligence but has had no training in the modern principles of dynamics. Above all, it is a physics that is particularly well adapted to the concepts of an earth at rest. Sometimes this is known as Aristotelian physics, because the major exposition of it in antiquity came from the philosopher-scientist Aristotle, who lived in Greece in the fourth century B.C. Aristotle was a pupil of Plato, and was himself tutor of Alexander the Great, who, like Aristotle, came from Macedonia.

ARISTOTLE'S PHYSICS OF COMMON SENSE

Aristotle was an important figure in the development of thought, and not for his contributions to science alone. His writings on politics and economics are masterpieces, and his works on ethics and metaphysics still challenge philosophers. Aristotle is looked upon as the founder of biology; Charles Darwin paid him this homage a hundred years ago: "Cuvier and Linnaeus have both been in many ways my two gods, but neither of them could hold a candle to old Aristotle." It was Aristotle who first introduced the concept of classification of animals, and he also brought to a high point the method of controlled observation in the biological sciences. One subject he studied was the embryology of the chick; it was his ambition to discover the sequence of development of the organs. Methodically he opened fertilized chicks' eggs on successive days, and made careful comparisons to find out the stages whereby the chick develops from an unformed embryo to a perfectly formed young chicken. Aristotle also was the first to formalize the process of deductive reasoning, in the form of the syllogism:

All men are mortal.
Socrates is a man.
Therefore, Socrates is mortal.

Aristotle pointed out that what makes such a set of three statements a valid progression is not the particular content of "man," "Socrates," and "mortal," but rather the form. For another example: all minerals are heavy, iron is a mineral, therefore iron is heavy. This is one of many valid forms of syllogism that were described by Aristotle in his great treatise on logic and reasoning, comprising both deduction and a form of induction.

Aristotle also stressed the importance of observation in the sciences, notably astronomy. For instance, among the many arguments he advanced to prove that the earth is more or less a sphere was the shape of the shadow cast by the earth on the moon, as observed during an eclipse. If the earth is a sphere, then the shadow cast by the earth is a cone; thus when the moon enters the earth's shadow, the shape of the shadow will always be roughly circular.

The importance of observation may be seen clearly in Aristotle's description of the moon rainbow:

> The rainbow is seen by day, and it was formerly thought that it never appeared by night as a moon rainbow. This opinion was due to the rarity of the occurrence; it was not observed, for, though it does happen, it does so rarely. The reason is that the colors are not easy to see in the dark and that many other conditions must coincide, and all that in a single day in the month. For if there is to be a moon rainbow it must be at full moon, and then as the moon is either rising or setting. So we have met with only two instances of a moon rainbow in more than fifty years.

These examples are sufficient to show that Aristotle cannot be described as purely an "armchair philosopher." It is true, nevertheless, that Aristotle did not put every statement to the test of experiment. He undoubtedly believed what he had been told by his teachers, just as successive generations believed what Aristotle had said. Often this is taken to be a basis for criticizing both Aristotle and his successors as scientists. But it should be kept in mind that students never verify all the statements they read, or even most of them, especially those found in textbooks or handbooks. Life is too short.

THE "NATURAL" MOTION OF OBJECTS

Now let us examine Aristotle's statements about motion. Basic to Aristotle's discussion was the principle that all the objects we encounter on this earth are made up of "four elements," air, earth, fire, and water. These are the "elements" we talk about in ordinary conversation when we say that someone out in a storm has "braved the elements." We mean that such a person has been in a windstorm, a dust storm, a rainstorm, and so on, not that he has struggled through a tornado of pure hydrogen or fluorine. Aristotle observed that some objects on earth appear to be light and others appear to be heavy. He attributed the property of being heavy or light to the proportion in each body of the different elements—earth being "naturally" heavy and fire being "naturally" light, water and air being intermediate between those two extremes. What, he asked, is the "natural" motion of such an object? He replied that if it is heavy, its natural motion will be downward, whereas if it is light its natural motion will be upward. Smoke, being light, ascends straight upward unless blown by the wind, while a rock, an apple, or a piece of iron falls straight downward when dropped. Hence, for Aristotle, the "natural" (or unimpeded) motion of a terrestrial object is straight upward or straight downward, upward and downward being reckoned along a straight line from the center of the earth through the observer.

Aristotle was, of course, aware that very often objects move in ways other than those just described. For instance, an arrow shot from a bow starts its flight apparently in a straight line that is more or less perpendicular to a line from the center of the earth to the observer. A ball at the end of a string can be whirled around in a circle. A rock can be thrown straight upward. Such motion, according to Aristotle, is "violent" or contrary to the nature of the body. Such motion occurs only when some force is acting to start and to keep the body moving contrary to its nature. A rock with a string tied around it can be lifted upward, and so made to undergo violent motion, but the moment the string is broken the rock will begin to fall downward in a natural motion, seeking its natural place.

Let us now consider the motion of heavenly objects: the stars, the planets, and the sun itself. These bodies appear to move around the earth in circles—the sun, moon, planets, and stars rising in the east, traveling through the heavens, and setting in the west (except for those circumpolar stars which move in small circles but never get below the horizon). According to Aristotle, the celestial bodies are not made of the same four elements as the earthly bodies. They are made of a "fifth element" or "aether." The natural motion of a body composed of aether is circular, so that the observed circular motion of the heavenly bodies is their natural motion, according to their nature, just as motion upward or downward in a straight line is the natural motion for a terrestrial object.

THE "INCORRUPTIBLE" HEAVENS

In the Aristotelian philosophy the heavenly bodies have one or two other properties of interest. The aether of which they are made is a material which is unchangeable, or to use the old word "incorruptible." This is in contrast to the four elements we find on earth—they are subject to change, that is, they are "corruptible." Thus on the earth we find both "coming into being" and "decay" and "passing away," things being born and dying. But in the heavens nothing ever changes; all remains the same: the same stars, the same eternal planets, the same sun, the same moon. The planets, the stars, and the sun were considered to be "perfect" and throughout the centuries were often compared to eternal diamonds or precious stones because of their unchanging qualities. The only heavenly object in which any kind of change or "imperfection" could be detected was the moon—but the moon, after all, is the heavenly body nearest the earth, and was considered a kind of dividing point between the terrestrial region of change (or corruptibility) and the celestial region of permanence and incorruptibility.

It should be observed that in this system all the heavenly objects circling the earth are more or less alike, and are all different from the earth—in physical characteristics, composition, and "essential properties." Thus one might understand why the earth remains still and does not move, whereas all the heavenly objects do move. Furthermore, the earth not only was said to have no "local motion," or movement from one place to another, but was not even supposed to rotate upon its axis. The chief physical reason for this, according to the old system, was that it is not "natural" for the earth to have a circular motion; that would be contrary to its nature, whether motion in orbit around the sun or a daily rotation upon its own axis.

THE FACTORS OF MOTION

Let us now examine a little more closely the Aristotelian physics of motion for terrestrial bodies. In all motion, said Aristotle, there are two major factors: the motive force, which we shall denote here by F, and the resistance, which we shall denote by R. For motion to occur, according to Aristotle, it is necessary that the motive force be greater than the resistance. Therefore our first principle of motion is

$$F > R \qquad [1]$$

or force must be greater than resistance. Let us next explore the effects of different resistances, all the while keeping the motive force constant. Our experiment will be performed with falling bodies, each of which will be allowed to fall freely, starting from rest, through a different resistant medium. In order to keep the conditions constant, we shall have the falling bodies all be spheres, so that the effect of their shape on their motion will be the same. Aristotle was, of course, quite aware that the speed of an object, all other things being equal, generally depends upon its shape, a fact we already have demonstrated with our parachute.

Now the experiment. Two identical steel balls of the same size, shape, and weight are used. We shall allow the two to fall simultaneously, one through air, the other through water. To do this experiment, you need a long cylinder filled with water; hold the two balls side by side, one over the water and the other at the same height but just outside this column of water (Fig. 1). When you release them simultaneously, you will see that there is no question that the speed of the one moving through air is very much greater than that of the one falling through the water. To prove that the results of the experiment did not derive from the fact that the balls were made of steel or had a particular weight, the experiment can be repeated using smaller steel balls, a pair of glass balls or brass balls, and so on. On a smaller scale, anyone can repeat this experiment by using two glass "marbles" and a highball glass filled to the brim with water. The result of this experiment can be written in the form of an equation, in which we express the fact that, all other things being equal, the speed in water (which greatly resists or impedes the motion) is less than

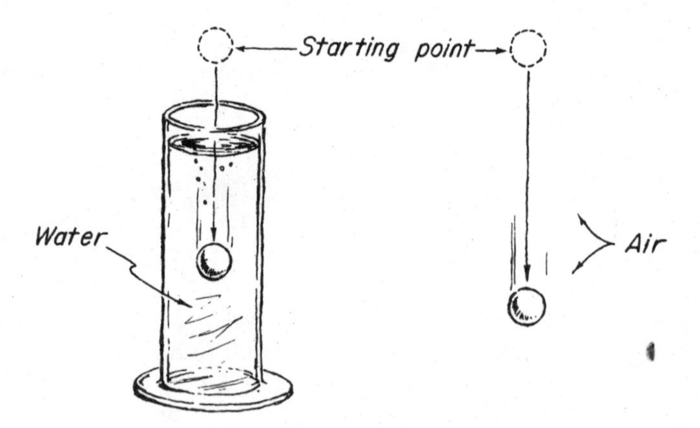

FIG. 1.

the speed in air (which does not impede the motion as much as water does):

$$V \propto \frac{1}{R} \qquad [2]$$

or the speed is inversely proportional to the resistance of the medium through which the body moves. It is a common experience that water resists motion; anyone who has tried to run through the water at the edge of the beach knows how much the water resists his motion in comparison to the air.

The experiment is now to be performed with two cylinders, one filled with water and the other filled with oil (Fig. 2). The oil resists the motion even more than the water; when the two identical steel spheres are dropped simultaneously, the one in water reaches the bottom long before the one falling through oil. Because the resistance R_o of oil is greater than the resistance R_w of water, we can now predict that if any pair of identical objects is let fall through these liquids, the one falling through water will drop through a given height faster than the one falling through oil. This prediction can easily be verified. Next, since it has been

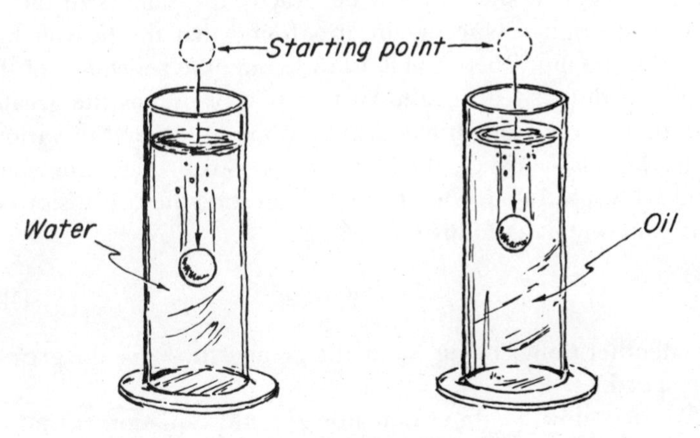

FIG. 2.

found that the resistance R_w of water is greater than the resistance R_a of air,

$$R_o > R_w$$
$$R_w > R_a \quad [3]$$

the resistance of oil must necessarily be greater than that of air,

$$R_o > R_a \quad [4]$$

This, too, can be verified by repeating the initial experiment with a cylinder filled with oil rather than water.

Let us next observe the effects of different motive forces. In this experiment we again use the long cylinder filled with water. In it we drop a small and a large steel ball simultaneously. We find that the large steel ball, the heavier of the two, reaches the bottom before the lighter one. Here, it might be argued, the size could have some effect, but if anything the larger ball should meet a greater resistance than the smaller one. Thus the experiment may serve to indicate that the greater the force to overcome a particular resistance, the greater the speed. This experiment may be repeated, this time using one ball of steel and the other of glass, so that the two will be exactly the same size but of different weights. Once again, it is found that the heavier ball seems to be much better able to overcome the resistance of the medium; thus it reaches the bottom first or attains the greater speed. The experiment can also be done in oil, and in various other liquids—alcohol, milk, and so on—to produce the same general result. In equation form, we can state the conclusions of this experiment as follows:

$$V \propto F \quad [5]$$

or, all other things being equal, the greater the force the greater the speed.

We may now combine Equation (2) and Equation (5) into a single equation as follows:

$$V \propto \frac{F}{R} \qquad [6]$$

or the speed is proportional to the motive force and inversely proportional to the resistance of the medium, or the speed is proportional to the force divided by the resistance. This equation is often known as the Aristotelian law of motion. It should be pointed out that Aristotle himself did not write his results in the form of equations, a modern way of expressing such relationships. Aristotle and most early scientists, including Galileo, preferred to compare speeds to speeds, forces to forces, and resistances to resistances. Thus instead of writing Equation (5) as we have done, they would have preferred the statement

$$V_g : V_s :: F_g : F_s$$

The ratio of speeds of the glass and steel balls is compared with the ratio of the forces with which these balls are moved downward. This is equivalent to the general statement that the speed of the glass ball is to the speed of the steel ball as the motive force of the glass ball is to the motive force of the steel ball.

Let us now study Equation (6), in order to discover some of its limitations. It is clear that this equation cannot be applied generally, because if the motive force should equal the resistance, the equation would not give the result that the speed V is equal to zero; nor does it give us a zero result when the force F is less than the resistance R. Hence Equation (6) is subject to the arbitrary limitation imposed by Equation (1), and is only true when the force is greater than the resistance. In other words, the equation is a limited and not a universal statement of the conditions of motion.

It is sometimes held that this equation may have arisen from the study of an unequal arm balance, say with equal weights on the two arms, or perhaps an equal arm balance with unequal weights at the ends of the two arms. In this case it is impossible for F to be less than R, because the greater weight is always the motive force, while the lesser weight is always the resistance.

Furthermore, in the equal arm balance if $F = R$ there will be no motion.

There are two final aspects of the law of motion, which we must introduce before we leave the subject. The first is that the law itself does not tell us anything about the stages by which an object falling from rest acquires the speed V. The law only tells us something about the speed itself: obviously some kind of "average" speed or "final" speed, since its measure is simply the time rate of traversing a given distance

$$V \propto \frac{D}{T} \qquad [7]$$

which holds for average speed or for motion at a constant speed, but not for accelerated or constantly changing speeds. Was it not known to Aristotle that the speed of a falling body starts from zero and by gradual stages attains its final value?

MOTION OF BODIES FALLING THROUGH AIR

Perhaps of greater significance to us than any of the preceding arguments is the outcome of another experiment. Thus far we have given the kind of positive experience that would make us have confidence in Aristotle's law of motion, but we have omitted one very crucial experiment. Let us return to a consideration of two objects of the same size, the same shape, but of different weight, or of different motive force F. We said that if these were dropped simultaneously through water, or through oil, it would be observed that the heavier one would descend more quickly. (The reader—before going on with the rest of this chapter and the rest of this book—will find it interesting to stop and perform these experiments for himself.) Now we come to the last in that earlier sequence of experiments; it consists of dropping two objects of the same size but of unequal weight in the same medium, but having the medium be *air.* Let us assume that the weight of one of our objects is exactly twice the weight of the other, which might imply in the old view that the speed of the heavier object should be just twice that of the lighter one. For a constant distance of fall, the speed is inversely proportional to the time, so that

$$V \propto \frac{1}{T} \qquad [8]$$

or

$$\frac{V_1}{V_2} = \frac{T_2}{T_1} \qquad [9]$$

or the speeds are inversely proportional to the times of descent. Hence, the time of descent of the heavier ball should be just half the time of descent of the lighter one. To perform the experiment, stand on a chair and drop the two objects together so that they will strike the bare floor. One good way of dropping them more or less simultaneously is to hold them horizontally between the first and second fingers of one hand. Then suddenly open the two fingers, and the two balls will begin to fall together. What is the result of this experiment?

Instead of describing the results of this experiment, let me suggest that you do it for yourself. Then compare your results with those obtained by John the Grammarian and also with the description given in the sixteenth century by Stevin, and finally with that given by Galileo in his famous book *Two New Sciences* some 350 years ago. (See pp. 6–8 above.) As John the Grammarian, Stevin, Galileo, and others easily found, the predictions of the Aristotelian theory are contradicted by experiment.*

One question you should ask yourself at this point is this: Evidently Equation (6) does not hold for air, but did it really hold for the other media which we explored? In order to see whether or not Equation (6) is an accurate quantitative statement, ask yourself whether it was merely a definition of "resistance," or, if

*For relatively short distances of fall, say from the ceiling of an ordinary room to the floor, the two balls will strike the ground with a single thud, unless there is a "starting error," an error arising from the fact that the two balls were not released simultaneously. A slight difference, such as Galileo and John the Grammarian observed, will occur for a greater distance of fall.

there is some other means of measuring "resistance," how the speeds were measured. Is it enough, in order to measure speed, to use Equation (8), and to measure the time of fall?*

In any event, most of you, I think, will have found that with the exception of the experiment of two unequal objects falling through air, the Aristotelian system sounds reasonable enough to be believed. There is no cause for us to condemn unduly either Aristotle or any Aristotelian physicist who had never performed the experiment of simultaneously dropping two objects of unequal weight in air.

THE IMPOSSIBILITY OF A MOVING EARTH

But what, you may still ask, has any of this to do with the earth's being at rest rather than in motion? For the answer let us turn to Aristotle's book *On the Heavens.* Here one finds the statement that some have considered the earth to be at rest, while others have said the earth moves. But there are many reasons why the earth cannot move. In order to have a rotation about an axis, each part of the earth would have to move in a circle, says Aristotle; but the study of the actual behavior of its parts shows that the

*We do not know how many scientists before Galileo and Stevin may have performed experiments with falling bodies. In an article on "Galileo and Early Experimentation" (in Rutherford Aris, H. Ted Davis, and Roger H. Stuewer, eds., *Springs of Scientific Creativity* [Minneapolis: University of Minnesota Press, 1983]), Thomas B. Settle describes such experiments performed by some Italians of the sixteenth century. Benedetto Varchi, a Florentine, wrote in a book of 1544 that "Aristotle and all other philosophers" never doubted, but "believed and affirmed" that the speed of a falling body is as its weight, but an experimental "test [*prova*] . . . shows it not to be true." It is not clear from the test whether Varchi had actually performed the experiment or was reporting an experiment made by others, Fra Francesco Beato and Luca Ghini. Giuseppe Moletti, a mathematician who had the same post of professor of mathematics at Pisa that Galileo later held, wrote a tract in 1576 in which he described how he had confuted Aristotle's conclusion, that in the motion of falling from a tower, a lead ball of 20 pounds will have a velocity 20 times that of a one-pounder. "They both arrive at the same time," Moletti wrote, "and I have made the test [*prova*] of it not once but many times." Moletti also made a test with balls of the same size, but of different materials (and hence having different weights), one of lead and one of wood. He found that when the two were released simultaneously from a high place, they would "descend and reach the ground or soil in the same moment of time."

natural earthly motion is along a straight line toward the center. "The motion, therefore, being enforced [violent] and unnatural, could not be eternal; but the order of the world is eternal." The natural motion of all bits of earthly matter is toward the center of the universe, which happens to coincide with the center of the earth. In "evidence" that earthly bodies do in fact move toward the center of the earth, Aristotle says, "We see that weights moving toward the earth do not move in parallel lines," but apparently at some angle to one another. "To our previous reasons," he then points out, "we may add that heavy objects, if thrown forcibly upwards in a straight line, come back to their starting place, even if the force hurls them to an unlimited distance." Thus, if a body were thrown straight up, and then fell straight down, these directions being reckoned with respect to the center of the universe, it would not land on earth exactly at the spot from which it was thrown, if the earth moved away during the interval. This is a direct consequence of the "natural" quality of straight-line motion for earthly objects.

The preceding arguments show how the Aristotelian principles of natural and violent (unnatural) motion may be applied to prove the impossibility of terrestrial movement. But what of the Aristotelian "law of motion," given in Equation (6) or Equation (9)? How is this specifically related to the earth's being at rest? The answer is given clearly in the beginning of Ptolemy's *Almagest,* the standard ancient work on geocentric astronomy. Ptolemy wrote, following Aristotelian principles, that if the earth had a motion "it would, as it was carried down, have got ahead of every other falling body, in virtue of its enormous excess of size, and the animals and all separate weights would have been left behind floating on the air, while the earth, for its part, at its great speed, would have fallen out of the universe itself." This follows plainly from the notion that bodies fall with speeds proportional to their respective weights. And many a scientist must have agreed with Ptolemy's final comment, "But indeed this sort of suggestion has only to be thought of in order to be seen to be utterly ridiculous."

CHAPTER
3

The Earth and the Universe

Very often the year 1543 is taken to be the natal year of modern science. In that year there were published two major books that led to significant changes in man's concept of nature and the world: one was the Polish churchman Nicholas Copernicus's *De revolutionibus orbium coelestium (On the Revolutions of the Celestial Spheres)* and the other, the Fleming Andreas Vesalius's *On the Fabric of the Human Body.* The latter dealt with man from the point of view of exact anatomical observation, and so reintroduced into physiology and medicine the spirit of empiricism that had characterized the writings of the Greek anatomists and physiologists, of whom the last and the greatest had been Galen. Copernicus's book introduced a new system of astronomy, which ran counter to the generally accepted notion that the earth is at rest. It will be our purpose here to discuss only certain selected features of the Copernican system, notably some consequences of considering the earth to be in motion. We shall not consider in any detail the relative advantages and disadvantages of the system as a whole, nor even compare its merits step by step with those of the older system. Our primary consideration is to explore what consequences the concept of a moving earth had for the development of a new science—dynamics.

COPERNICUS AND THE BIRTH OF MODERN SCIENCE

In ancient Greece it was suggested that the earth may have a daily rotation on its axis and make an annual revolution in a huge orbit around the sun. Proposed by Aristarchus in the third century

B.C., this system of the universe lost out to one in which the earth is at rest. There was great opposition to the idea that the earth can be in motion. Even when, almost 2000 years later, Copernicus published his account of a system of the universe based on a combination of the two terrestrial motions, there was no immediate assent. Eventually, of course, Copernicus's book proved to have contained the seed of the whole scientific revolution that culminated in Isaac Newton's magnificent foundation of modern physics. Looking backwards, we can see how the acceptance of the Copernican concept of a moving earth necessarily implied a non-Aristotelian physics. Was this sequence apparent to the contemporaries of Copernicus? And why did not Copernicus himself produce that scientific revolution which has altered the world to such an extent that we still are not fully aware of all its consequences? In this chapter we shall explore these questions, and in particular we shall see why Copernicus's proposal of a system of the world in which the earth is held to be in motion and the sun to be at rest was not of itself sufficient for a rejection of the old physics.

At the outset we must make it plain that Copernicus (1473–1543) was in many ways more of a conservative than a revolutionary. Many of the ideas he introduced had already existed in the literature, and again and again the fact that he was unable to go beyond the basic principles of Aristotelian physics hampered him. When we talk today about the "Copernican system," we usually mean a system of the universe quite different from that described in Copernicus's *De revolutionibus orbium coelestium.* The reason for this procedure is that we wish to honor Copernicus for his innovations, and do so at the expense of literal accuracy by referring to the sun-centered system of the post-Copernican era as "Copernican." It should more properly be called "Keplerian" or at least "Keplero-Copernican."

THE SYSTEM OF CONCENTRIC SPHERES

But before describing the Copernican system, let me state some of the basic features of the two principal pre-Copernican systems. One, attributed to Eudoxus, was improved by another Greek astronomer, Callippus, and received its finishing touches from Aristotle. This is the system known as the "concentric spheres." In this system each planet—and also the sun and the moon—was considered to be fixed on the equator of a separate sphere, which rotates on its axis, the earth being stationary at the center. While each sphere is rotating, the ends of the axis of rotation are fixed in another sphere, which is also rotating—with a different period and about an axis that does not have the same orientation as the axis of the inner sphere.

For some planets there could be as many as four spheres, each embedded in the next, with the result that there would be a variety of motions. For instance, one of these spheres could account for the fact that wherever the planet happened to be among the stars it would be made to revolve once around the earth in every 24 hours. There would be another such sphere to move the sun in its daily apparent revolution, another for the moon, and another for the fixed stars. The set of inner spheres for each planet would account for the fact that a planet does not appear to move through the heavens with only a daily motion, but also shifts its position from day to day with respect to the fixed stars. Thus a planet will sometimes be seen in one constellation, and again in another. Because planets are seen to wander among the fixed stars from night to night, they derive the name "planet" from the Greek verb meaning "to wander." One of the observed features of this wandering is that its direction is not constant. The usual direction of motion is to progress slowly eastward, but every now and again the planet stops its eastward motion (reaching a stationary point) and then (Fig. 3) moves for a short while westward, until it reaches another stationary point, after which it resumes its original eastward motion through the heavens. The eastward motion is known as "direct" motion, the westward motion as "retrograde." By the proper combination of spheres Eudoxus was able to construct a model to show how combinations of circular motion could produce the observed direct and retrograde apparent motion of the planets. It is somewhat the same kind of "spheres" that appear in the title of Copernicus's book.

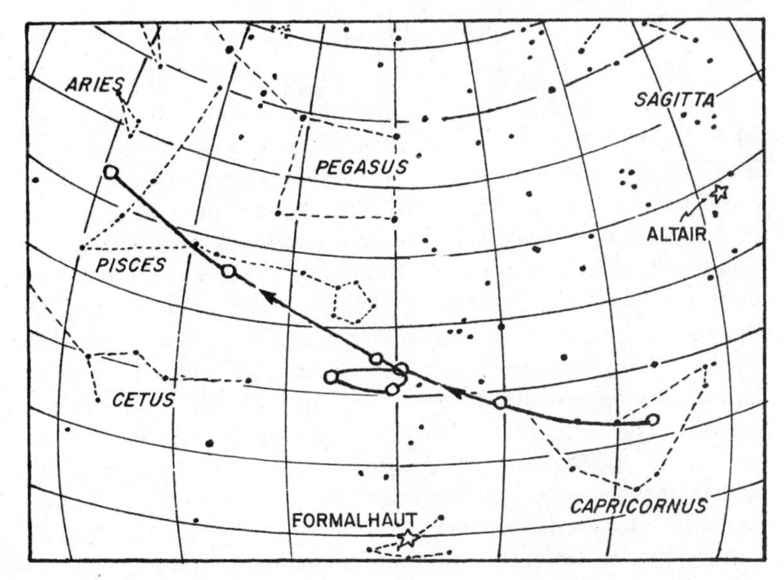

FIG. 3.

After the decline of Greece, science fell into the hands of the Islamic or Arabian astronomers. Some among them elaborated the system of Eudoxus and Aristotle and introduced many further spheres in order to make the predictions of this system agree more exactly with observation. These spheres, obtaining a certain reality, were even thought to be made of crystal; the system acquired the title of "crystalline spheres." Because it was held that the orientation of the stars and planets had an important influence on all human affairs, men and women came to believe that the influence of the planet emanates not from the object itself but from the sphere to which it is attached. In this belief we may see the origin of the expression "sphere of influence," still used today in a political and economic sense.

PTOLEMY AND THE SYSTEM OF EPICYCLES AND DEFERENTS

The other major rival system of antiquity was elaborated by Claudius Ptolemy, one of the greatest astronomers of the ancient world, and was based in some measure on concepts that had been introduced by the geometer Apollonius of Perga and the astronomer Hipparchus. The final product, known generally as the Ptolemaic system, in contrast to the Eudoxus-Aristotle system of homocentric (common-centered) spheres, had enormous flexibility, and as a consequence enormous complexity. The basic devices were used in various combinations. First of all, consider a point *P* moving uniformly in a circle around the point *E*, as in Fig. 4A. Here is an illustration of uniform circular motion that permits neither stationary points nor retrogradation. Nor does it account for the fact that the planets do not have a constant speed as they appear to move around the earth. At most such a motion could be observed only in the behavior of the fixed stars, for Hipparchus had seen even the sun moving with variable velocity, an observation connected with the fact that the seasons are not of the same length. In Fig. 4B, the earth is not at the exact center *C* of that circle, but is off-center, at the point *E*. Then it is clear that if the point *P* corresponds to a planet (or to the sun), it will

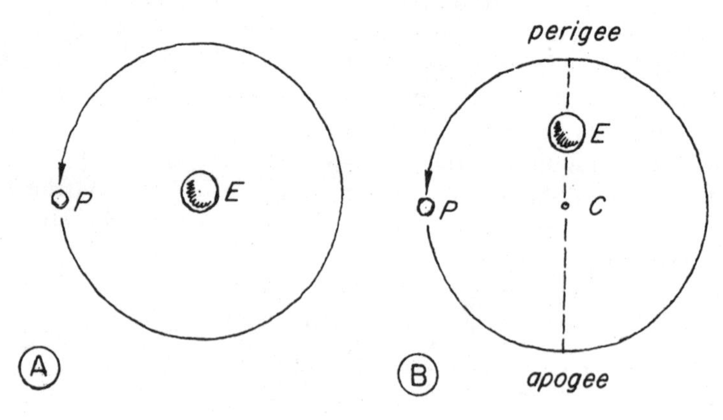

FIG. 4.

not appear to move uniformly with respect to the fixed stars as seen from the earth, even though its motion along the circle is in fact uniform. If the earth and heavenly body form such an eccentric system, rather than a homocentric system, there will be times when the sun or planet will be very near the earth (perigee), and times when the sun or planet will be very far from the earth (apogee). Thus we should expect a variation in the brightness of the planets, which is also observed.

Next, we shall introduce one of Ptolemy's chief devices to account for the motion of the planets. Let us assume that while the point P moves uniformly on a circle about the center C (Fig. 5), a second point Q moves in a circle about the point P. The result will be to produce a curve with a series of loops or cusps. The large circle on which P moves is called the circle of reference, or the deferent, and the small circle on which Q moves is called the epicycle. Thus the Ptolemaic system is often described as one based on deferent and epicycle. It is clear that the curve resulting from the combination of epicycle and deferent is one in which the planet at some times is nearer the center than at others, that there are also stationary points, and that when the planet is on the inside of each loop, an observer at C will see it move with a retrograde motion. In order to make the motion conform to observation, it is necessary only to choose the relative size of epicycle and deferent, and the relative speeds of rotation of the two circles, so as to conform to the appearances.

It is plain from his book that Ptolemy did not ever commit himself on the question whether there were "real" epicycles and "real" deferents in the heavens. As a matter of fact, it seems much more likely that for him the system that he described was a "model" of the universe, and not necessarily the "true" picture —whatever those words may mean. That is, it was the Greek ideal, reaching its highest point in the writings of Ptolemy, to construct a model that would enable the astronomer to predict the observations, or—to use the Greek expression—"to save the appearances." Although often disparaged, this approach to science is very similar to that of the twentieth-century physicist,

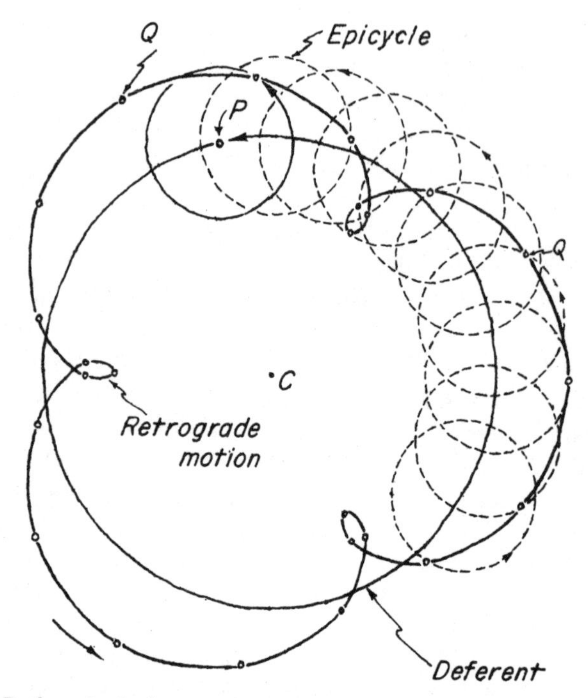

FIG. 5. Ptolemy's device to account for the wanderings of the planets assumed a complicated combination of motions. The planet Q traveled around P in a circle (dotted lines) while P moved in a circle around C. The solid line with loops is the path Q would follow in the combined motion.

whose primary aim is also to produce a model that will yield equations predicting the results of experiment. Often today's physicist must be satisfied with equations in the absence of a "model" in the ordinary workable sense.

Certain other features of the old Ptolemaic system may be listed briefly. The earth need not be at the center of the deferent circle, or, expressed differently, the deferent circle (Fig. 6A) could be eccentric rather than homocentric—that is, with a center different from the center of the earth. Furthermore, while the point *P* is moving about the big circle (Fig. 6B) of reference or

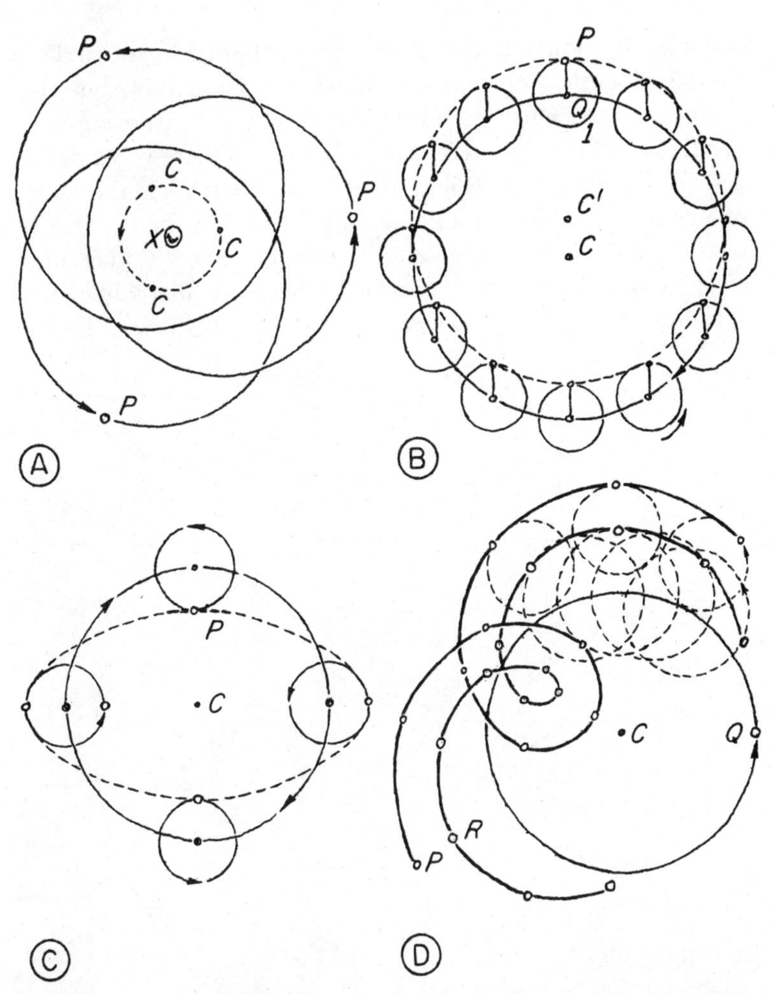

FIG. 6. With epicycle and deferent (and ingenuity) astronomers could describe almost any observed planetary motion and still stay within the bounds of the Ptolemaic system. In (A) point P moves on circle with center C, which moves on smaller circle centered at X. In (B) the effect of the combination of deferent and epicycle is to shift the apparent center of P's orbit from C to C′. In (C) the combination yields an elliptical curve. The figure in (D) traces the path of P moving along an epicycle on an epicycle; the center of P's circle is R, which moves on a circle whose center, Q, is on a circle centered at C.

deferent, its center *C* could be moving about a small circle, a combination that need not produce retrogradation, but that could have the effect of lifting the circle or transposing it or producing elliptical motion (Fig. 6C). Finally, there was a device known as the "equant" (Fig. 7). This was a point not at the center of a circle about which motion could be "uniformized." That is, consider a point *P* moving on a circle with center at *C* in relation to an equant. The point *P* moves in such a way that a line from *P* to the equant sweeps out equal angles in equal times; this has

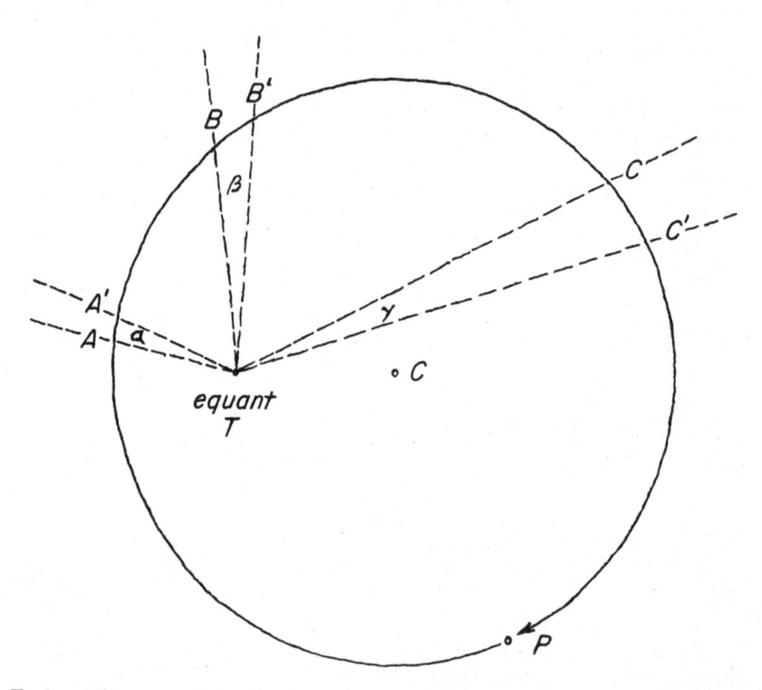

FIG. 7. The equant was a Ptolemaic device to explain apparent changes in a planet's speed. While the movement of P from A to A′, from B to B′, and from C to C′ would not be uniform with respect to the center of the circle, C, it would be with respect to another point, T, the equant, because the angles α, β, γ are equal. The planet moves along each of the arcs AA′, BB′, and CC′ in the same time but, obviously, at different speeds.

the effect that P does not move uniformly along its circular path for an observer elsewhere than at the equant. These devices could be used in many different combinations. The result was a system of much complexity. Many a man of learning could not believe that a system of forty or more "wheels within wheels" could possibly be turning about in the heavens, that the world was so complicated. It is said that Alfonso X, King of León and Castile, called Alfonso the Wise, who sponsored a famous set of astronomical tables in the thirteenth century, could not believe the system of the universe to be that intricate. When first taught the Ptolemaic system, he commented, according to legend: "If the Lord Almighty had consulted me before embarking upon the creation, I should have recommended something simpler."

Nowhere have the difficulties of understanding the Ptolemaic system been expressed so clearly as by the poet John Milton in his famous poem *Paradise Lost.* Milton had been a schoolteacher, had actually taught the Ptolemaic system, and knew, therefore, whereof he wrote. In these lines the angel Raphael is replying to Adam's questions about the construction of the universe and telling him that God must surely be moved to laughter by men's activities:

> *. . . when they come to model Heav'n*
> *And calculate the Stars, how they will wield*
> *The mighty frame, how build, unbuild, contrive*
> *To save appearances, how gird the Sphere*
> *With Centric and Eccentric scribbled o'er,*
> *Cycle and Epicycle, Orb in Orb . . .*

Before we go into the innovations of Copernicus, a few final remarks on the old system of astronomy may be in order. In the first place, it is clear that part of the complexity arose from the fact that the curves representing the apparent motions of the planets (Fig. 5) are combinations of circles. If one could simply have used an equation for a cusped curve such as a lemniscate, the job would have been a great deal simpler. One must keep in mind, however, that in Ptolemy's day there was no analytic geometry using equations, and that a tradition had grown up, sanctioned by both Aristotle and Plato, that the motion of the heavenly bodies must be explained in terms of a natural system of motion—perhaps on the argument that a circular motion has neither beginning nor end and is therefore most fitting to the unchangeable, incorruptible, ever-moving planets. In any event, as we shall see, the idea of explaining planetary motion solely by combinations of circles remained in astronomy for a long, long time.

The Ptolemaic system not only worked or could be made to

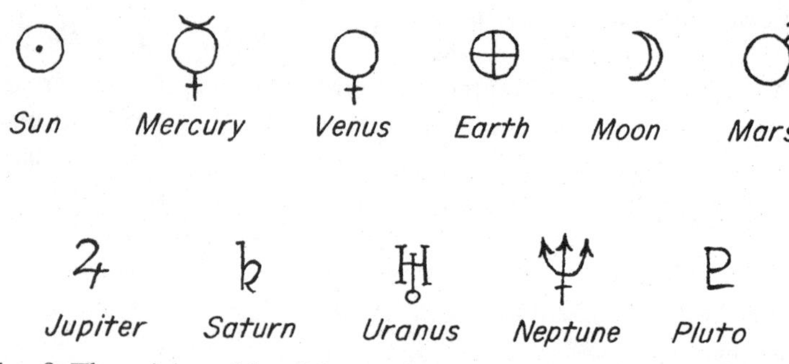

FIG. 8. The origins of the oldest planetary symbols are lost in antiquity, but the commonly accepted derivations stem from Latin and Greek mythology. The sun symbol probably represented a buckler (shield) with boss. The symbol for Mercury represented either his caduceus, the staff he carried, or his head and winged cap. The symbol of Venus was the looking glass associated with the goddess of love and beauty. The symbol for Mars, god of war, has been taken to represent either a warrior's head and helmet with nodding plume or a spear and shield. The symbol for Jupiter also has alternative derivations—either a crude hieroglyph of the eagle, "bird of Jove," or the first letter of Zeus, the Greek name of Jupiter. Saturn's symbol is an ancient scythe, emblem of the god of time. The symbol for Uranus is the first letter of its discoverer's name, Sir William Herschel (1738–1822), with the planet suspended from the crossbar. The trident was always carried by Neptune, god of the sea. The symbol for Pluto is an obvious monogram. It is interesting that the alchemists used the Mercury symbol for the metal mercury and the Venus symbol for copper. Today geneticists designate female with the Venus symbol and male with the Mars symbol.

work, but fitted in perfectly with the system of Aristotelian physics. The stars, planets, sun, and moon were assigned motions in circles or in combinations of circles, their "natural motion," while the earth did not partake of motion, being in its "natural place" at the center of the universe, and at rest. In the Ptolemaic system, then, there was no need to seek a new system of physics other than the one that accorded equally well with the system of homocentric spheres. Sometimes these two systems are described as being "geostatic," because in both of them the earth is at rest; the more customary expression is "geocentric," because in both of the systems the earth is at the center of the universe.

COPERNICAN INNOVATIONS

As Copernicus elaborated his own system, it bore many resemblances to the system of Ptolemy. Copernicus admired Ptolemy enormously; in organizing his book, ordering the different chapters and choosing the sequence in which various topics were introduced, he followed Ptolemy's *Almagest.*

The transfer from a geostatic to a heliostatic (immobile sun) system did involve certain new explanations. To see them, let us begin as Copernicus did by first considering the simplest form of the heliostatic universe. The sun is at the center, fixed and immobile, and around it there move in circles in the following order: Mercury, Venus, the earth with its moon, Mars, Jupiter, Saturn (Fig. 8A). Copernicus explained the daily apparent motions of the sun, moon, stars, and planets on the ground that the earth rotates upon its axis once a day. The other major appearances derived, he said, from a second motion of the earth, which was an orbital revolution about the sun, just like the orbits of the other planets. Each planet has a different period of revolution, the period being greater the farther the planet is from the sun. Thus retrograde motion is easily explained. Consider Mars (Fig. 9), which moves more slowly around the sun than the earth. Seven positions of the earth and Mars are shown at a time when

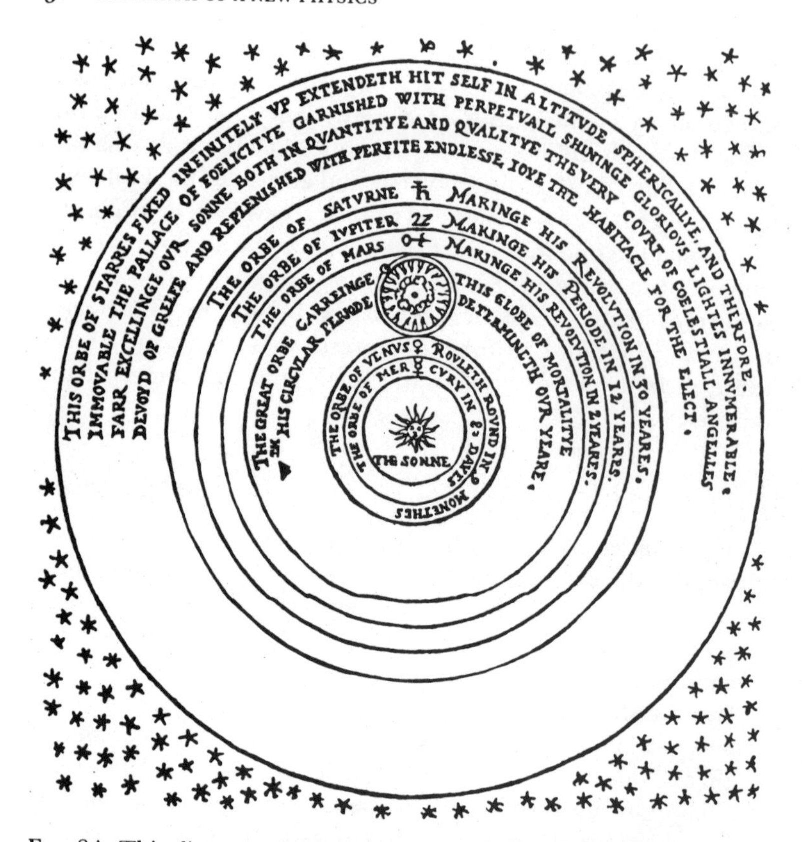

FIG. 8A. This diagram of the Copernican system is taken from Thomas Digges's *A Perfit Description of the Caelestial Orbes* (1576), giving an English translation of a portion of Copernicus's *De revolutionibus.* Digges has added one feature to the system in making the sphere of the fixed stars infinite.

the earth is passing Mars and when Mars is in opposition (that is, when a line from the sun to Mars passes through the earth). It will be seen that a line drawn from the earth to Mars at each of these successive positions will move first forward, then backward, and then forward again. Thus Copernicus not only could explain "naturally" how retrograde motion occurs, but also could show why it is that retrogradation is observed in Mars only at opposi-

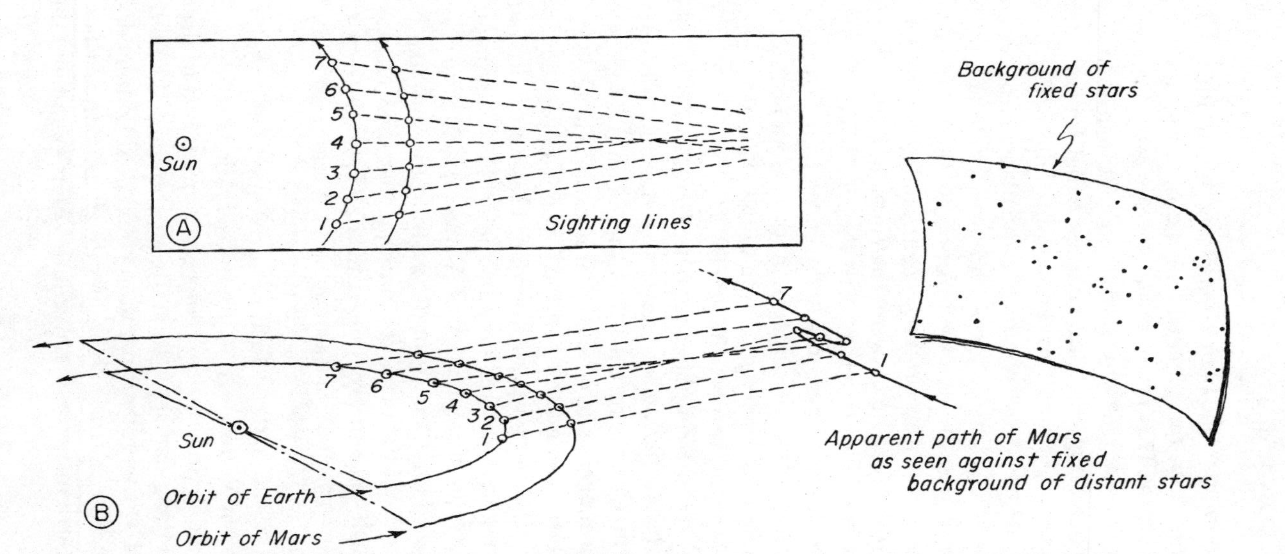

FIG. 9. In the Copernican system the apparent retrograde motion of planets has a simple explanation; it is a matter of relative speeds. Here the sighting lines show why a superior planet, one farther from the sun than the earth is, seems to reverse its direction. It is traveling around the sun more slowly than the earth is.

tion, corresponding to the planet's crossing the meridian at midnight. In opposition, the planet is on the opposite side of the earth from the sun. This is why it will reach its highest position in the heavens at midnight, or will cross the meridian at midnight. In similar fashion (Fig. 10) one could see that for an inferior planet (Mercury or Venus) retrogradation would occur only at inferior conjunction, corresponding to the planet's crossing the meridian at noon. (When Venus or Mercury lies along a straight line from the earth to the sun, the position is called conjunction. These planets are in the center of retrogradations at inferior conjunction, when they lie between the earth and the sun. Then they cross the meridian together with the sun at noon.) These two facts make perfect sense in a heliocentric or heliostatic system, but if the *earth* were the center of motion, as in the Ptolemaic system, why should the retrogradation of the planets depend on their orientation with respect to the *sun?*

Continuing with the simplified model of circular orbits, let us observe next that Copernicus was able to determine the scale of the solar system. Consider Venus (Fig. 11). Venus is seen only as evening star or morning star, because it is either a little ahead of the sun or a little behind the sun but never 180 degrees away from the sun, as a superior planet may be. The Ptolemaic system (Fig. 11A) accounted for this only by the arbitrary assumption that the centers of the epicycles of Venus and Mercury were permanently fixed on a line from the earth to the sun; that is to say, the deferents of Mercury and Venus, just like the sun, moved around the earth once in every year. In the Copernican system one had merely to assume that the orbits of Venus and Mercury (Fig. 11B) were within the orbit of the earth.

In the Copernican system, furthermore, one could compute the distance from Venus to the sun. Observations made night after night would indicate when Venus could be seen at its greatest elongation (angular separation) from the sun. When this event occurred, the angular separation could be determined. As may be seen in Fig. 12, the maximum elongation occurs when a line from the earth to Venus is tangent to Venus's orbit and thus

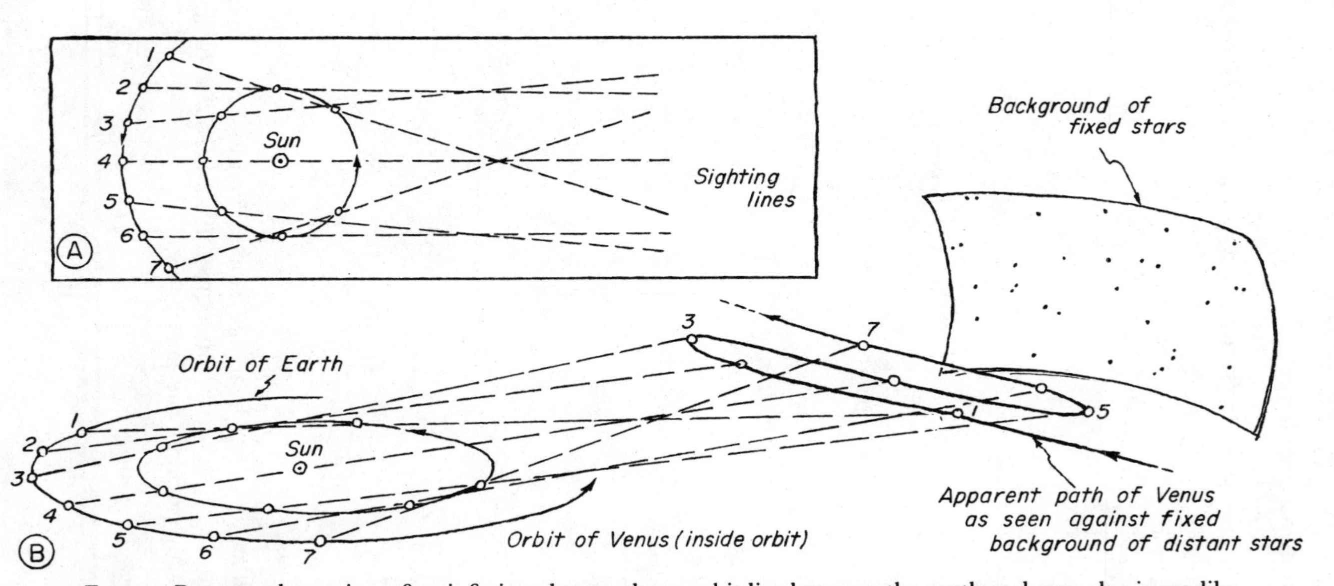

FIG. 10. Retrograde motion of an inferior planet, whose orbit lies between the earth and sun, also is readily explained with sighting lines. Venus travels around the sun faster than the earth does.

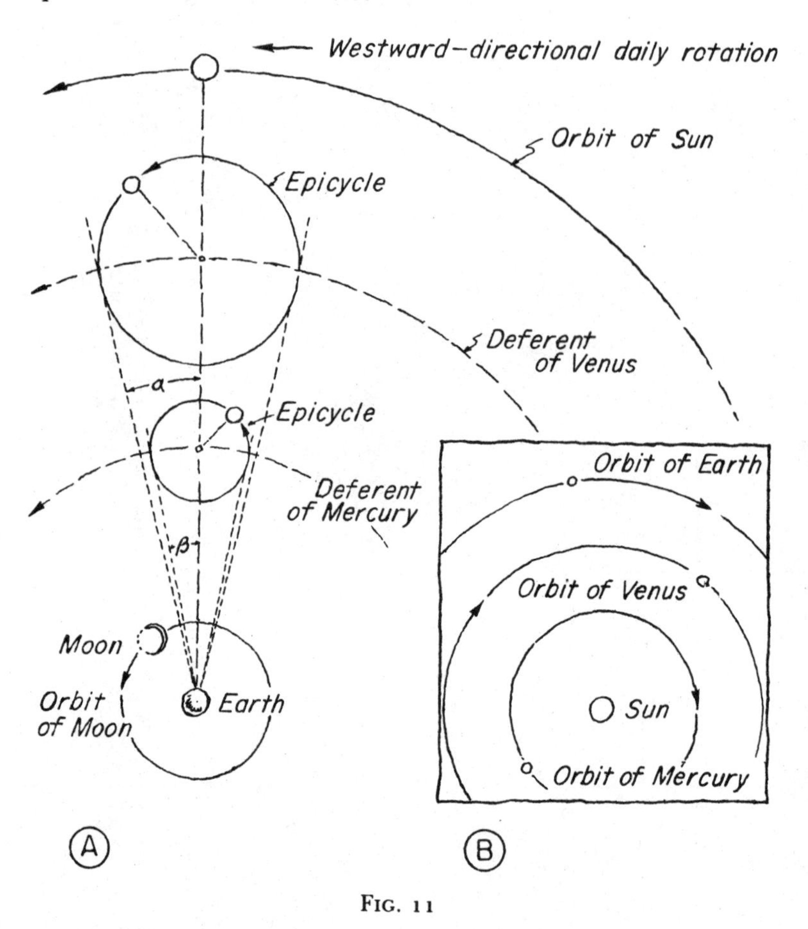

FIG. 11

perpendicular to a line from the sun to Venus. From simple trigonometry we can write this equation and from a table of tangents easily calculate the length VS.

$$\frac{VS}{ES} = \text{sine } \alpha \qquad [1]$$

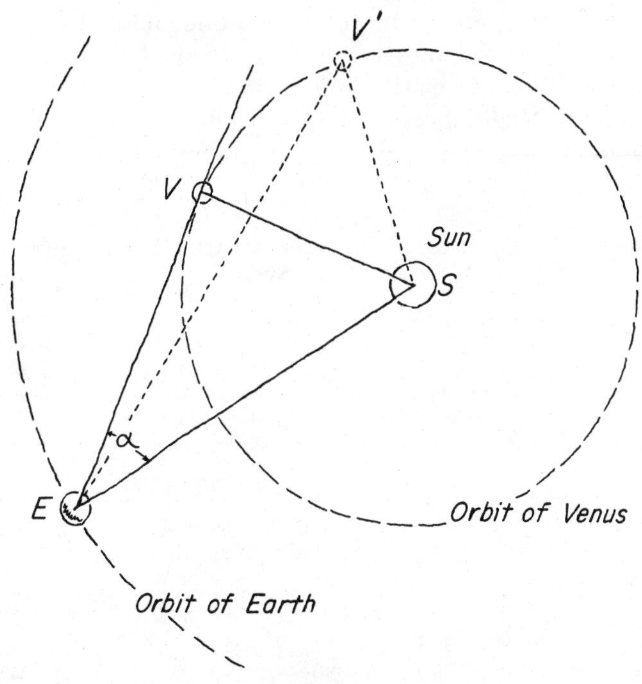

FIG. 12. Computing the distance from Venus to the sun became possible in the Copernican system. When the angular separation (that is, the angle α of Venus from the sun) is at the maximum, the line of sight from the earth to Venus (EV) is tangent to Venus's orbit and therefore perpendicular to the radius VS. Computing the length of VS is an easy problem in elementary trigonometry. At any other orientation, say V', the angular separation is not maximum.

The distance ES, or the average size of the radius of the earth's orbit in the Copernican system, is known as an "astronomical unit." Thus Equation (1) may be rewritten as

$$\text{VS} = (\text{sine } \alpha) \times 1\text{AU} \qquad [2]$$

By the use of this simple method Copernicus was able to determine the planetary distances (in astronomical units) with great accuracy, as may be seen from the following table, which shows Copernicus's values and the present accepted values for the planetary distances from the sun. (The Copernican method for determining the distances from the sun differs slightly in the case of the three "superior" planets: Mars, Jupiter, Saturn.)

COMPARISON OF COPERNICAN AND MODERN VALUES FOR THE ELEMENTS OF THE SOLAR SYSTEM

	Mean Synodic Period*		*Sidereal Period*		*Mean Distance from Sun***	
Planet	C	M	C	M	C	M
Mercury	116d	116d	88d	87.91d	0.36	0.391
Venus	584d	584d	225d	225.00d	0.72	0.721
Earth			365¼d	365.26d	1.0	1.000
Mars	780d	780d	687d	686.98d	1.5	1.52
Jupiter	399d	399d	12y	11.86y	5	5.2
Saturn	378d	378d	30y	29.51y	9	9.5

*Synodic periods are times between conjunctions of the same bodies.
**Expressed in astronomical units.

Furthermore, Copernicus was able to determine with equal accuracy the time required for each planet to complete a revolution of 360 degrees around the sun, or its sidereal period. Since Copernicus knew the relative sizes of the planetary orbits and the sidereal periods of the planets, he was then able to predict to a tolerable degree of accuracy the planets' future positions (that is, their respective distances from the earth). In the Ptolemaic system, the distances of the planets played no role whatsoever, since there was no way of determining them from observations. So long as the relative sizes and relative periods of motion on deferent and epicycle were the same, the observations or appearances would be identical, as may be seen in Fig. 13. That the Ptolemaic system dealt chiefly in angle rather than in distance may be seen most clearly in the example of the moon. It was one of the major features of the Ptolemaic system that the moon's apparent position could be described with a relatively high degree of accuracy. But this required a special device, and had the moon really followed the contrived path it would have had an enormous variation in apparent size, far greater than is observed. Until recent

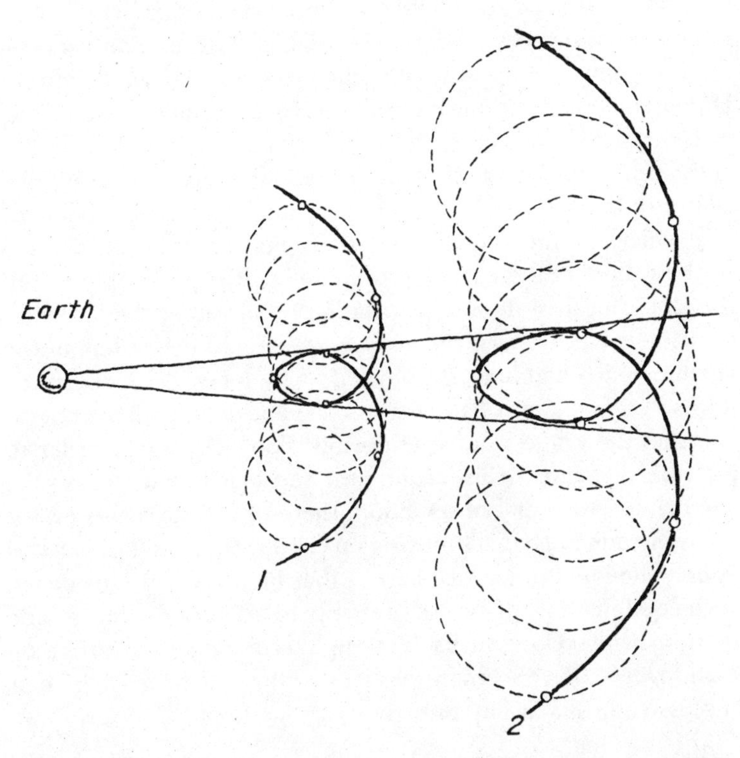

FIG. 13. In the Ptolemaic system predictions of planetary positions leaned on measurement of angles, not distances. This illustration shows that observations would be the same regardless of distance if the relative periods of motion were the same.

years, it was believed that Copernicus's own theory of the moon was one of his most original innovations. But we now know that the identical theory existed in Islamic astronomy.

I have said earlier that the system of a single circle for each planet with a single circle for the moon, and two different motions for the earth, constitutes a simplified version of the Copernican system. The fact of the matter is that such a system does not agree with observation, except in a rough way. In order to make his system more accurate, therefore, Copernicus found it necessary to introduce a number of complexities, many of which remind us of devices used in the Ptolemaic system. For instance, it was obvious to Copernicus (as the inverse had been obvious to Hipparchus) that the earth cannot move uniformly about a circle with the sun at the center. Thus Copernicus placed the sun not at the center of the earth's orbit, but at some distance away. The center of the solar system, and of the universe, in the system of Copernicus is thus not the sun at all, but rather a "mean sun," or the center of the earth's orbit. Hence, it is preferable to call the Copernican system a heliostatic system rather than a heliocentric system. Copernicus objected greatly to the system of the equant, which had been introduced by Ptolemy. For Copernicus it was necessary, as it had been for the ancient Greek astronomers, that the planets move uniformly along circles. In order to produce planetary orbits around the sun that would give results conforming to actual observation, therefore, Copernicus ended up by introducing circles moving on circles, much as Ptolemy had done. The chief difference here is that Ptolemy had introduced such combinations of circles primarily to account for retrograde motion, while Copernicus (Fig. 14) accounted for retrograde motion, as we have seen, by the fact that the planets move in their successive orbits at different speeds.* A comparison of the two

*A final complexity of the Copernican system arose from the difficulties Copernicus experienced in accounting for the fact that the axis of the rotating earth remains fixed in its orientation with respect to the stars even though the earth moves in its orbit. The "motion" introduced by Copernicus was found to be unnecessary. Galileo later showed that because no force is acting to turn the earth's axis, it does *not* move but always remains parallel to itself.

figures representing the Ptolemaic and Copernican systems does not show that one was in any obvious way "simpler" than the other.

COPERNICUS VERSUS PTOLEMY

What were the advantages and disadvantages of the Copernican system as compared to the Ptolemaic system? In the first place, one decided advantage of the Copernican system was the relative ease in explaining retrograde motion of planets and showing why their positions relative to the sun determined the retrograde motions. A second advantage of the Copernican system was that it afforded a basis for determining the distances of the planets from the sun and from the earth.

It is sometimes said that the Copernican system was a great simplification, but this is based upon a misunderstanding. If the Copernican system is considered in the rudimentary form of a single circle for each planet around the sun, then this assumption is valid. But such a system of pure and simple circles can only be a crude approximation, as Copernicus knew well. We have seen that in order to obtain a more accurate representation of the planetary motions, he had recourse to a combination of circle moving on circle, somewhat reminiscent of Ptolemy's epicyclic constructions, though for a different purpose.

Let us next explore the reasons for not accepting the Copernican system. A major one was the absence of any annual parallax of the fixed stars. The phenomenon of parallax is the shift in view that occurs when the same object is seen from two different positions. This is the principle upon which range-finders for artillery and for photographic cameras are built. Consider the motion of the earth in the Copernican system. If the stars are examined at intervals six months apart, this is equivalent to making observations from the ends of a base line almost 200 million miles long (Fig. 15), because the radius of the earth's orbit around the sun is 93 million miles. Since Copernicus and the astronomers of his day could not determine any parallax of the fixed stars by such semiannual observations, it had to be assumed that the stars are enormously far off, if indeed the earth does move around the sun.

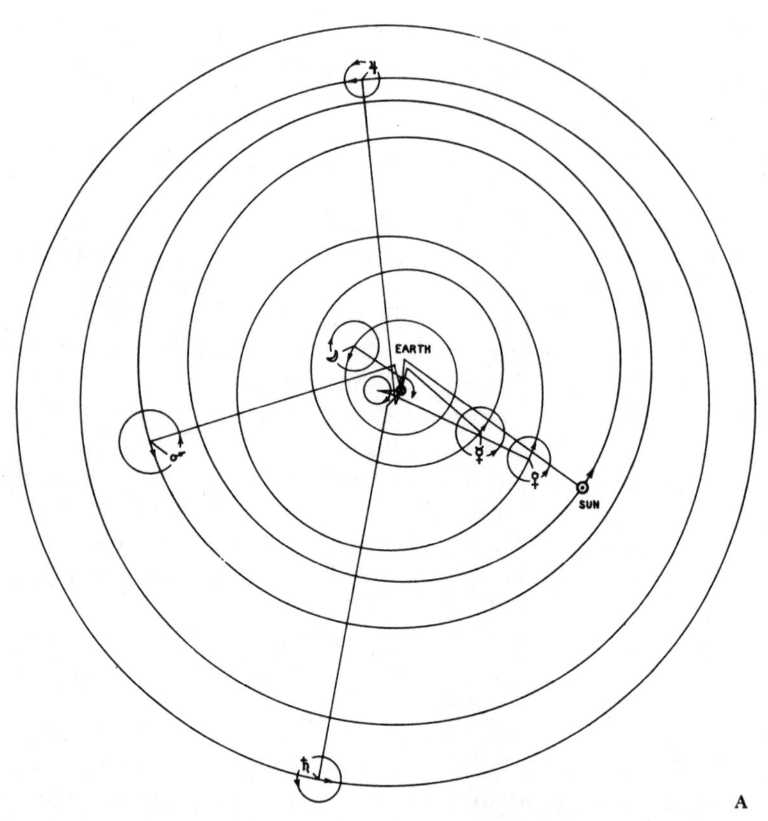

A

Fig. 14. The Ptolemaic system (A) and the Copernican system (B) were of about equal complexity, as can be seen in this comparison. The dots at the inner ends of the radii of the planets' deferents (large circles) denote the centers of the orbits relative to the center of the sun's orbit in the Ptolemaic system and relative to the sun in the Copernican system. Note the use of epicycles in both systems. In this diagram, the centers of the epicycles of Venus (♀) and of Mercury (☿) have been displaced for greater visibility. In the Ptolemaic system, the centers of these two epicycles remain fixed on a straight line drawn from the earth to the sun. (After William D. Stahlman)

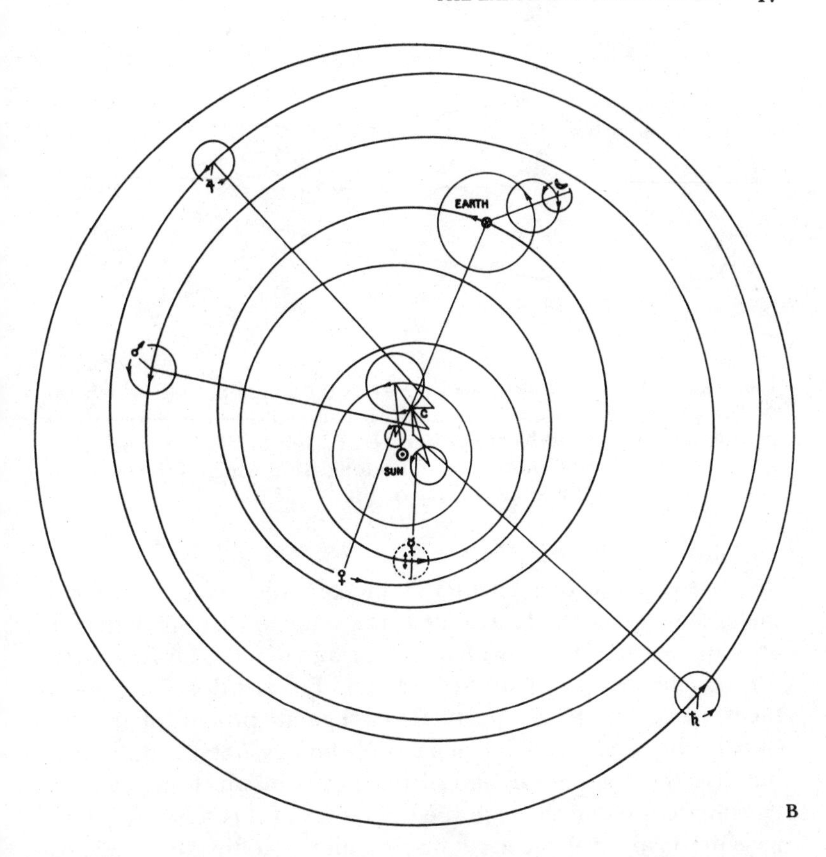

B

It was far simpler to say that the absence of any observed annual parallax of the fixed stars tended to disprove the whole basis of the Copernican system. Many centuries after Copernicus, in fact about 150 years ago, greatly improved telescopes permitted astronomers to observe just such a parallax of the fixed stars. Until that time, however, the existence of such a parallax (which had to be very small) had to be accepted by astronomers as a matter of faith.

From the failure of astronomical observation, let us turn next to the failure of mechanics. How did Copernicus explain the motion of bodies on a moving earth? These are the problems we

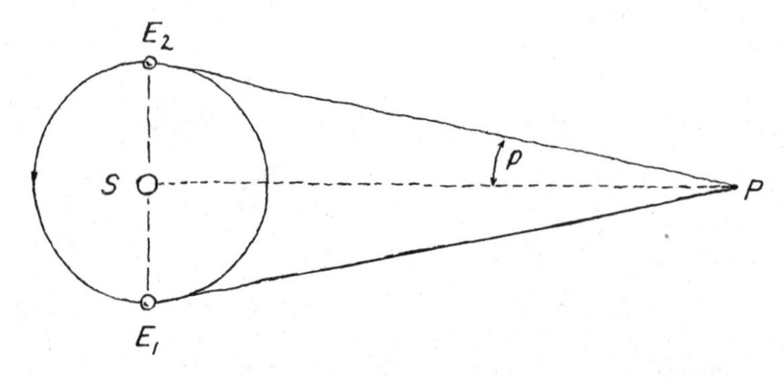

FIG. 15. The annual parallax of a star is the angle p, with which the distance from the sun and earth can be calculated. The earth's positions at intervals of six months are designated E_1 and E_2. The distance E_1E_2 gives a base line of 200,000,000 miles long from which to observe the star P and obtain the angle E_1PE_2, or 2p.

discussed in the first chapter, none of which Copernicus explained adequately. He assumed that somehow or other the air around the earth moves with the earth, and that this air is in some way attached to the earth. According to Edward Rosen, "Copernicus's theory of gravity postulated a separate process of gravitational cohesion for individual heavenly bodies, not only the earth but also the sun, moon, and planets, each of which maintained its spherical shape through the operation of this tendency. Objects in the air near the earth may be subject to this tendency, or the nearby air and the objects in it may share in the earth's rotation because they are contiguous with it. In offering these alternative suggestions (*Revolutions* I, 8–9), Copernicus made germinal contributions to what later developed into the concepts of universal gravitation and inertia."

But there was another problem, in some ways even more difficult to account for—the nature of the solar system itself. If Copernicus still held to the principles of Aristotelian physics—and he never invented a new physics to take the place of the Aristotelian—how could he explain that the earth seems to move in a daily rotation and in an annual circular orbit, both of them contrary to its nature? In point of fact, Copernicus was forced to say that the earth moving around the sun is "merely another planet." But to say that the earth is "merely another planet" must have seemed a denial of the Aristotelian principle that the earth and the planets are made of different materials, are subject to different sets of physical laws, and behave therefore in different ways. For the earth to move in a circular orbit about the sun might appear to imply that the earth was undergoing violent motion; but Aristotelian physics attributed a natural linear motion only to objects made of earthly matter, and not to the earth as a whole. In the old Aristotelian physics, in fact, the earth could properly have no motion at all, neither natural motion nor violent motion. Copernicus argued that, in general, "rotation is natural to a sphere"; and thus he was led (*Revolutions* I, 8) to conclude that—since the earth has a spherical shape—"if anyone believes that the earth rotates, surely he will hold that its motion is natural, not violent." While Copernicus was thereby introducing an extension of Aristotelian physics that actually contradicts the basic precepts of Aristotle (as that the earth cannot move), he did not elaborate a fully workable new system of physics adequate to the range of problems posed by conceiving the earth to be in motion.

Many who have read Copernicus's book must have been puzzled by his statement that the earth necessarily has a rotation about its axis as well as a motion in a large circle around the sun, that this follows from the fact that the earth has a spherical shape. As we have seen, Copernicus argued that it is the "nature" of a sphere to be in spherical motion. How, then, could Copernicus also assert that the sun, which has a spherical shape, stands still and neither rotates about its axis nor moves in an annual revolution?

One final problem of a physical nature that Copernicus had to cope with involves the moon. In the Copernican system, it could be explained that while the earth moves around the sun, falling objects continue to fall straight down, and birds are not lost, because the air is somehow or other glued to the earth. That is, Copernicus (*Revolutions* I, 8) supposed that because the air around the earth is somehow or other "linked" to the earth, it shares in the earth's motions; that is, it rotates with the earth and moves along in space with the orbiting planet. Hence, as the earth rotates on its axis and moves in orbit around the sun, the air causes falling objects to keep their position relative to the earth while they fall, so that—to a terrestrial observer—they appear to be falling straight downward. They have a motion that is, accordingly, "twofold; being in every case a compound of straight and circular." Copernicus does not discuss the argument about birds and other living creatures or even clouds, but the case is much the same as for the rising and falling of bodies. But this argument cannot be extended to the moon, since Copernicus held that only the air relatively near the earth is carried along with the earth. If we go far out, away from the earth, we reach "that part of the air" which, Copernicus maintains, "is unaffected by the earth's motion on account of its great distance from the earth." Some other explanation is required for the moon. Here was a question that was difficult for Copernicus to answer.

Thus far we have confined our attention to two aspects of the Copernican system: the fact that it was at least as complex as the Ptolemaic system, and the fact that apparently insoluble problems of physics arose if one accepted his system. If we add to these objections some other general difficulties in the Copernican system, it may readily be seen that publication of his book in 1543 could not of itself achieve a revolution in physical or astronomical thought.

PROBLEMS WITH A COPERNICAN UNIVERSE

Apart from the purely scientific problems, the concept of a moving earth created serious intellectual challenges. When all is said and done, it *is* rather comforting to think that our abode is fixed in space and has a proper place in the scheme of things, rather than being an insignificant speck whirling aimlessly somewhere or other in a vast and perhaps even infinite universe. The Aristotelian uniqueness of the earth, based on its supposedly fixed position, gave people a sense of pride that could hardly arise from being on a rather small planet (compared to Jupiter or Saturn) in a rather insignificant location (position 3 out of 7 successive planetary orbits). To say the earth is "merely another planet" suggests that it may not have even the distinction of being the only inhabited globe, and this implies that earthly man himself is not unique. And perhaps other stars are suns with other planets and on each are other kinds of men and women. Most people of the sixteenth century were not ready for such views, and the evidence of their senses reinforced their bias. Planet indeed! Anyone who looks at a planet—Venus, Mars, Jupiter, or Saturn—will "see" at once that it is "another star" and not "another earth." The fact that these planetary "stars" are brighter than the others, wander with respect to the others, and may have an occasional retrograde motion does not make them different from the other (or fixed) stars; such properties "obviously" do not make the "wandering stars" (which we call planets) in any way like this earth on which we stand. And if it were not enough that all "common sense" rebels at the idea of the earth as "merely another planet," there is the evidence of Scripture. Again and again Holy Writ mentions a moving sun and a fixed earth. Even before the publication of *De revolutionibus,* Martin Luther heard about Copernicus's ideas and condemned them violently for contradicting the Bible. And everyone is well aware that Galileo's subsequent advocacy of the new system brought him into conflict with the Roman Inquisition.

It should be clear, therefore, that the alteration of the frame of the universe proposed by Copernicus could not be accomplished without shaking the whole structure of science and of our thought about ourselves. Copernicus's book finally led to a ferment in the thinking about the nature of the universe, and about the earth, that would eventually produce profound change. This is the sense in which we can date the first step of the scientific revolution at 1543. The problems posed and their implications penetrated the very foundations of physics and astronomy. From what has been said thus far, the way in which changes in one section of physical science affect the whole body of science should be clear. Practicing scientists today are familiar with this phenomenon, having witnessed the growth of modern atomic physics and quantum theory. Yet nowhere can the unity of the structure of science be seen better than in the fact that the Copernican system, whether in its simple or complex form, could not stand by itself as expounded by Copernicus. It required a modification of the currently held ideas about the nature of matter, the nature of the planets, the sun, the moon, and the stars, and the nature and actions of force in relation to motion. It has been well said that the significance of Copernicus lay not so much in the system he propounded as in the fact that the system he did propound would ignite the great revolution in physics that we associate with the names of such scientists as Galileo, Johannes Kepler, and Isaac Newton. The so-called Copernican revolution was really a later revolution of Galileo, Kepler, and Newton.

CHAPTER
4

Exploring the Depths of the Universe

The march of science has rhythms not wholly unlike those of music. As in sonatas, certain themes recur in a more or less orderly sequence of variations. The place of Copernicus in the history of science may well illustrate this process. Although his system was neither so simple nor so revolutionary as it often is represented, his book raised all the questions that had been lurking behind every cosmological scheme since antiquity. The elaborate proofs that Aristotle and Ptolemy had given of the immobility of the earth could never fully conceal from any reader that another view was possible, the one that Aristotle and Ptolemy had attacked.

EVOLUTION OF THE NEW PHYSICS

As in any well-structured musical composition, the main Copernican theme appears in separate parts. One man in antiquity, Heraclides of Pontus, had presented the concept of a rotation of the earth, but not an orbital motion, while Aristarchus had a scheme in which the earth both rotated on its axis and revolved around the sun as the planets do. In the Latin Middle Ages prior to Copernicus it was not uncommon to find thinkers like the Frenchman Nicole Oresme and the German Nicolaus Cusanus considering a possible motion (a motion of rotation) of the earth, and it would have been extraordinary indeed if the theme of the moving earth did not manifest itself again after Copernicus. *De revolutionibus* contained the most complete account of a heliostatic universe that had ever been composed, and for the specialist in astronomy and the cosmologist it proposed much that was new and important. In the same sense that the logic of a sonata leads from the original statement of a theme through successive variations, but does not dictate exactly what the variations shall be, so the logic of the development of science enables us to predict what some of the consequences of Copernicus's ideas would have to be, what changes in thought would necessarily follow on the acceptance of the new world view. But only a knowledge of history itself reveals that the gradual acceptance of Copernican ideas by one scholar here and another there was rudely interrupted in 1609, when a new scientific instrument changed the level and the tone of discussion of the Copernican and Ptolemaic systems to such a degree that the year overshadows 1543 in the development of modern astronomy.

It was in 1609 that scientists first began to use the telescope to make systematic studies of the heavens. The revelations proved that Ptolemy made specific errors and important ones, that the Copernican system neatly fitted the new facts of observation, and that the moon and the planets have properties making them very much like the earth in a variety of different ways and patently unlike the stars.

After 1609 any discussion of the respective merits of the two great systems of the world was bound to turn on phenomena that were beyond the ken and even the imagination of either Ptolemy or Copernicus. And once the heliocentric system was seen to have a possible basis in "reality," it would spur the search for a physics that would apply with equal validity on a moving earth and throughout the universe. The introduction of the telescope would have been enough by itself to turn the course of science, but another development of 1609 further accelerated the revolution: Johannes Kepler published his *Astronomia nova,* which not only simplified the Copernican system by getting rid of all the epicycles but also firmly established two laws of planetary motion, as we shall see in a later chapter.

GALILEO GALILEI

The scientist who was chiefly responsible for introducing the telescope as a scientific instrument, and who laid the foundations of the new observational astronomy and a new physics, was Galileo Galilei. In 1609 he was a professor at the University of Padua, in the Venetian Republic, and was forty-five years old, which is considerably beyond the age when people are usually held to make profoundly significant scientific discoveries. The last great Italian, except for nobles and kings, to be known to posterity by his first name, Galileo was born in Pisa, Italy, in 1564, almost on the day of Michelangelo's death and within a year of Shakespeare's birth. His father sent him to the University at Pisa, where his sardonic combativeness quickly won him the nickname "wrangler." Although his first thought had been to study medicine—it was better paid than most professions—he soon found that it was not the career for him. He discovered the beauty of mathematics and thereafter devoted his life to this subject, along with physics and astronomy. We do not know exactly when or how he became a Copernican, but on his own testimony it happened earlier than 1597.

Galileo made his first contribution to astronomy before he ever used a telescope. In 1604 a "nova" or new star suddenly appeared in the constellation Serpentarius. Galileo showed this to be a "true" star, located out in the celestial spaces and not inside the sphere of the moon. That is, Galileo found that this new star had no measurable parallax and so was very far from the earth. Thus he delivered a nice blow to the Aristotelian system of physics because he proved that change *could* occur in the heavens despite Aristotle, who had held the heavens unchangeable and had limited the region where change may occur to the earth and its surroundings. Galileo's proof seemed to him all the more decisive in that it was the second nova that observers found to have no measurable parallax. The previous one of 1572, in the constellation Cassiopeia, had been studied by the Danish astronomer Tycho Brahe (1546–1601), the major figure in astronomical science between Copernicus and Galileo. Among the achievements of Tycho were the design and construction of improved naked-eye instruments and the establishment of new standards of accuracy in astronomical observation. Tycho's nova, rivaling the brightness of Venus at its peak and then gradually fading away, shone for sixteen months. This star had no detectable parallax, and also did not partake of planetary motion, but remained in a constant orientation with respect to the other fixed stars. Tycho correctly concluded that change may occur in the region of the fixed stars no matter what Aristotle or any of his followers had said. Tycho's observations contributed to the cumulative evidence against Aristotle, but the crushing blow had to await the night when Galileo first turned his telescope to the stars.

THE TELESCOPE: A GIANT STEP

The history of the telescope is itself an interesting subject. Some scholars have attempted to establish that such an instrument had been devised in the Middle Ages. An instrument possibly like a telescope was described in a book published by Thomas Digges in 1571, and a telescope with an inscription stating that it had been made in Italy in 1590 was in the possession of a Dutch scientist around 1604. What effect, if any, these early instruments had on the ultimate development of telescopes we do not know; perhaps this is an example of an invention made and then lost again. But in 1608, the instrument was re-invented in Holland, and there are at least three claimants to the honor of having then made the "first" one. Who actually deserves the credit is of little concern to us here, because our main problem is to learn how the telescope changed the course of scientific thought. Sometime early in 1609 Galileo heard a report of the telescope, but without any specific information as to the way in which the instrument was constructed. He has recorded how:

> . . . A report reached my ears that a certain Fleming had constructed a spyglass by means of which visible objects, though very distant from the eye of the observer, were distinctly seen as if nearby. Of this truly remarkable effect several experiences were related, to which some persons gave credence while others denied them. A few days later the

> report was confirmed to me in a letter from a noble Frenchman at Paris, Jacques Badovere [a former pupil of Galileo], which caused me to apply myself wholeheartedly to inquire into the means by which I might arrive at the invention of a similar instrument. This I did shortly afterwards, my basis being the theory of refraction. First I prepared a tube of lead, at the ends of which I fitted two glass lenses, both plane on one side while on the other side one was spherically convex and the other concave. Then placing my eye near the concave lens I perceived objects satisfactorily large and near, for they appeared three times closer and nine times larger than when seen with the naked eye alone. Next I constructed another one, more accurate, which represented objects as enlarged more than sixty times. Finally, sparing neither labor nor expense, I succeeded in constructing for myself so excellent an instrument that objects seen by means of it appeared nearly one thousand times larger and over thirty times closer than when regarded with our natural vision.

Galileo was not the only observer to point the new instrument toward the heavens. It is even possible that two observers—Thomas Harriot in England and Simon Marius in Germany—were in some respects ahead of him. But there seems to be general agreement that the credit of first using the telescope for astronomical purposes may be given to Galileo and that this attribution is justified by "the persistent way in which he examined object after object, whenever there seemed any reasonable prospect of results following, by the energy and acuteness with which he followed up each clue, by the independence of mind with which he interpreted his observations, and above all by the insight with which he realized their astronomical importance," as said Arthur Berry, British historian of astronomy. Furthermore, Galileo was the first to publish an account of the universe as seen through a telescope. The "message" that Galileo disseminated throughout the world in his book of 1610 revolutionized astronomy. (See Supplement 1.)

It is impossible to exaggerate the effects of the telescopic discoveries on Galileo's life, so profound were they. Not only is this true of Galileo's personal life and thought, but it is equally true of their influence on the history of scientific thought. Galileo had the experience of beholding the heavens as they actually are for perhaps the first time,* and wherever he looked he found evidence to support the Copernican system against the Ptolemaic, or at least to weaken the authority of the ancients. This shattering experience—of observing the depths of the universe, of being the first mortal to know and to inform the world what the heavens are actually like—made so deep an impression upon Galileo that it is only by considering the events of 1609 in their proper proportion that one can understand the subsequent direction of his life. And it is only in this way that we can appreciate how there came about that great revolution in the science of dynamics that may properly be said to mark the beginning of modern physics.

To see the way in which these events occurred, let us turn to Galileo's account of his discoveries, in a book which he called *Sidereus nuncius,* that is, *The Sidereal Messenger* (which can also be translated as *The Starry Messenger* or *The Starry Message*). In its subtitle, the book is said to reveal "great, unusual, and remarkable spectacles, opening these to the consideration of every man, and especially of philosophers and astronomers." The newly observed phenomena, the title page of the book declared, were to be found "in the surface of the moon, in innumerable fixed stars, in nebulae, and above all in four planets swiftly revolving about Jupiter at differing distances and periods, and known to no one before the Author recently perceived them and decided that they should be named the Medicean Stars."

THE LANDSCAPE OF THE MOON

Immediately after describing the construction and use of the telescope, Galileo turned to results. He would "review the observations made during the past two months, once more inviting the attention of all who are eager for true philosophy to the first steps of such important contemplations."

The first celestial body to be studied was the moon, the most prominent object in the heavens (except for the sun), and the one

*He could not have known whether, in fact, any other observers had anticipated his study of the heavens through a telescope.

nearest to us. The crude woodcuts accompanying Galileo's text cannot convey the sense of wonder and delight this new picture of the moon awoke in him. The lunar landscape, seen through the telescope (Plates 2 and 3), unfolds itself to us as a dead world —a world without color, and so far as the eye can tell, one without any life upon it. But the characteristic that stands out most clearly in photographs, and that so impressed Galileo in 1609, is the fact that the moon's surface appears to be a kind of ghostly *earthly* landscape. No one who looks at these photographs, and no one who looks through a telescope, can escape the feeling that the moon is a miniature earth, however dead it may appear, and that there are on it mountains and valleys, oceans, and seas with islands in them. To this day, we refer to those oceanlike regions as "maria" even though we know, as Galileo later discovered, that there is no water on the moon, and that these are not true seas at all. (See Supplement 2.)

The spots on the moon, whatever may have been said about them before 1609, were seen by Galileo in a coldly new and different light (Plate 4). He found "that the surface of the moon is not smooth, uniform, and precisely spherical as a great number of philosophers believe it (and the other heavenly bodies) to be, but is uneven, rough, and full of cavities and prominences, being not unlike the face of the earth, relieved by chains of mountains and deep valleys." Galileo's brilliant style in describing the earth-like quality of the moon is apparent in the following extract:

> Again, not only are the boundaries of shadow and light in the moon seen to be uneven and wavy, but still more astonishingly many bright points appear within the darkened portion of the moon, completely divided and separated from the illuminated part and at a considerable distance from it. After a time these gradually increase in size and brightness, and an hour or two later they become joined with the rest of the lighted part which has now increased in size. Meanwhile more and more peaks shoot up as if sprouting now here, now there, lighting up within the shadowed portion; these become larger, and finally they too are united with that same luminous surface which extends further. And on the earth, before the rising of the sun, are not the highest peaks of the mountains illuminated by the sun's rays while the plains remain in shadow? Does not the light go on spreading while the larger

central parts of these mountains are becoming illuminated? And when the sun has finally risen, does not the illumination of plains and hills finally become one? But on the moon the variety of elevations and depressions appears to surpass in every way the roughness of the terrestrial surface, as we shall demonstrate further on.

Not only did Galileo describe the appearance of mountains on the moon; he also measured their height.* It is characteristic of Galileo as a scientist of the modern school that as soon as he found any kind of phenomenon he wanted to measure it. It is all very well to be told that the telescope discloses that there are mountains on the moon, just as there are mountains on the earth. But how much more extraordinary it is, and how much more convincing, to be told that there are mountains on the moon and that they are exactly four miles high! Galileo's determination of the height of the mountains on the moon has withstood the test of time, and today we agree with his estimate of their maximum height. (For those who are interested, Galileo's method of computing the height of these mountains will be found in Fig. 16.)

To see what a world of difference there is between Galileo's realistic description of the moon, which resembles the description that an aviator might give of the earth as seen from the air, and the generally prevailing view, read the following lines from Dante's *Divine Comedy.* Written in the fourteenth century, this work is generally considered to be the ultimate expression of the culture of the Middle Ages. In this part of the poem Dante has arrived on the moon and discusses certain features of it with Beatrice, who speaks to him with the "divine voice." This is how the moon appeared to this medieval space traveler:

Meseemed a cloud enveloped us, shining, dense, firm and polished, like diamond smitten by the sun.
Within itself the eternal pearl received us, as water doth receive a ray of light, though still itself uncleft. . . .

*It has been one of the marvels of our age that our astronauts have actually voyaged to the moon and have observed its surface to be as Galileo described it —a feat made visible to millions of observers on their television screens and recorded for posterity in the evidence of photographs and rock samples.

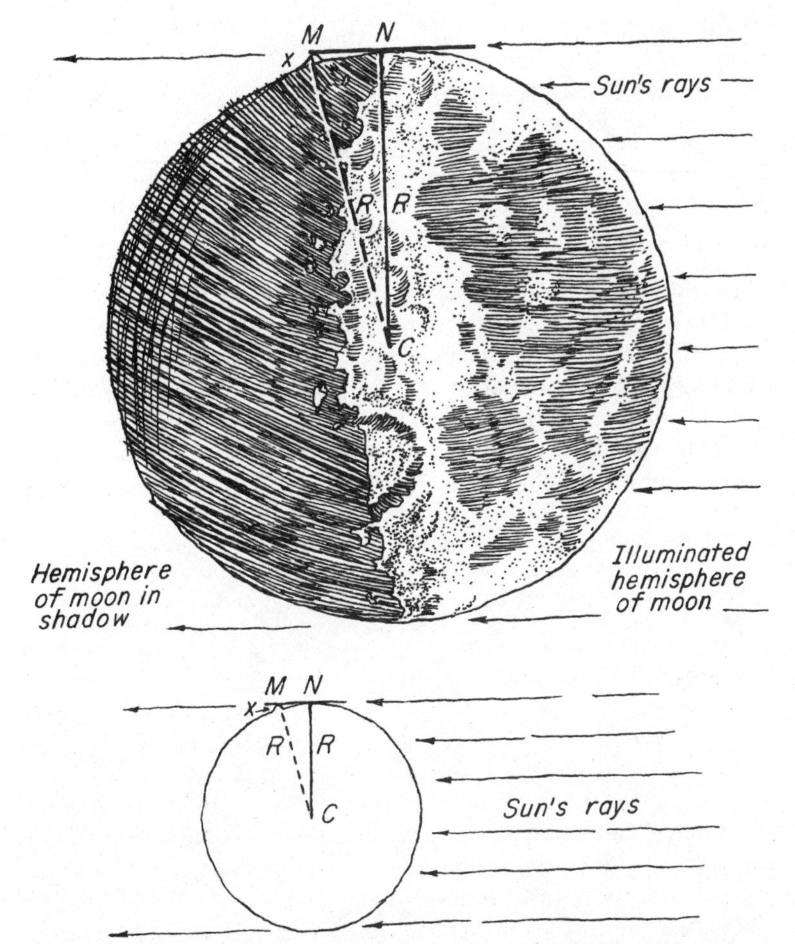

FIG. 16. Galileo's measurement of the height of mountains on the moon was simple but convincing. The point N is the terminator (boundary) between the illuminated and non-illuminated portions of moon. The point M is a bright spot observed in the shadowed region; Galileo correctly surmised that the bright spot was a mountain peak whose base remained shadowed by the curvature of the moon. He could compute the moon's radius from the moon's known distance from the earth and could estimate the distance NM through his telescope. By the Pythagorean theorem, then, $CM^2 = MN^2 + CN^2$, or, since R is the radius and x the altitude of the peak,

$$(R + x)^2 = R^2 + MN^2, \text{ or}$$
$$R^2 + 2Rx + x^2 = R^2 + MN^2, \text{ or}$$
$$x^2 + 2Rx - MN^2 = 0$$

which is easily solved for x, the altitude of the peak.

Dante asked Beatrice:

"But tell me what those dusky marks upon this body, which down there on earth make folk to tell the tale of Cain?"
She smiled a little, and then: "And if," she said, "the opinion of mortals goeth wrong, where the key of sense doth not unlock,
truly the shafts of wonder should no longer pierce thee; since even when the senses give the lead thou see'st reason hath wings too short. . . ."

Dante had written that man's senses deceive him, that the moon is really eternal and perfect and absolutely spherical, and even homogeneous. One should not overestimate the power of reason, he believed, since the human mind is not powerful enough to fathom the cosmic mysteries. Galileo, on the other hand, trusted the revelation of the senses enlarged by the telescope, and he concluded:

> Hence if anyone wished to revive the old Pythagorean opinion that the moon is like another earth, its brighter part might very fitly represent the surface of the land and its darker region that of the water. I have never doubted that if our globe were seen from afar when flooded with sunlight, the land regions would appear brighter and the watery regions darker. . . .

Apart from the statement about water, which Galileo later corrected, what is important in this conclusion is that Galileo saw that the surface of the moon provides evidence that the earth is not unique. Since the moon resembles the earth, he had demonstrated that at least the nearest heavenly body does not enjoy that smooth spherical perfection attributed to all heavenly bodies by the classic authorities. Nor did Galileo make this only a passing reference; he returned to the idea later in the book when he compared a portion of the moon to a specific region on earth: "In the center of the moon there is a cavity larger than all the rest, and perfectly round in shape. . . . As to light and shade, it offers the same appearance as would a region like Bohemia if that were enclosed on all sides by very lofty mountains arranged exactly in a circle."

EARTHSHINE

At this point Galileo introduces a still more startling discovery: earthshine. This phenomenon may be seen in the photograph reproduced in Plate 5. From the photograph it is plain, as may be seen when the moon is examined through a telescope, that there is what Galileo called a "secondary" illumination of the dark surface of the moon, which can be shown geometrically to accord perfectly with light from the sun reflected by the earth into the moon's darkened regions. It cannot be the moon's own light, or a contribution of starlight, since it would then be displayed during eclipses; it is not. Nor can it come from Venus or from any other planetary source. As for the moon's being illuminated by the earth, what, asked Galileo, is there so remarkable about this? "The earth, in fair and grateful exchange, pays back to the moon an illumination similar to that which it receives from her throughout nearly all the darkest gloom of night." However startling this discovery may have appeared to Galileo's readers, it must be noted that earthshine had previously been discussed by Kepler's teacher, Michael Mästlin, in a disputation on eclipses (1596), and by Kepler himself in his treatise of 1604 on optics.

Galileo ends his description of the moon by telling his readers that he will discuss this topic more fully in his book on the *System of the World.* "In that book," he said, "by a multitude of arguments and experiences, the solar reflection from the earth will be shown to be quite real—against those who argue that the earth must be excluded from the dancing whirl of stars [or heavenly bodies] for the specific reason that it is devoid of motion and of light. We shall prove the earth to be a wandering body [i.e., a planet] surpassing the moon in splendor, and not the sink of all dull refuse of the universe; this we shall support by an infinitude of arguments drawn from nature." This was Galileo's first announcement that he was writing a book on the system of the world, a work which was delayed for many years and which—when finally published—resulted in Galileo's trial before the Roman Inquisition and his condemnation and subsequent imprisonment.

But observe what Galileo had proved thus far. He showed that the ancients were wrong in their descriptions of the moon; the moon is not the perfect body they pictured, but resembles the earth, which therefore cannot be said to be unique and consequently different from all the heavenly objects. And if this was not enough, his studies of the moon had shown that the earth shines. No longer was it valid to say that the earth is not a shining object like the planets. And if the earth shines just as the moon does, perhaps the planets may also shine in the very same manner by reflecting light from the sun! Remember, in 1609 it was still an undecided question whether the planets shine from internal light, like the sun and the stars, or whether by reflected light, like the moon. As we shall see in a moment, it was one of Galileo's greatest discoveries that the planets shine by reflected light as they encircle the sun in their orbits.

STARS GALORE

But before turning to that subject, let us state briefly some of Galileo's other discoveries. When Galileo looked at the fixed stars, he found that they, like the planets, "appear not to be enlarged by the telescope in the same proportion as that in which it magnifies other objects, and even the moon itself." Furthermore, Galileo called attention to "the differences between the appearance of the planets and of the fixed stars" in the telescope. "The planets show their globes perfectly round and definitely bounded, looking like little moons, spherical and flooded all over with light; the fixed stars are never seen to be bounded by a circular periphery, but have rather the aspect of blazes whose rays vibrate about them and scintillate a great deal." Here was the basis of one of Galileo's great answers to the detractors of Copernicus. Plainly, the stars must be at enormous distances from the earth compared to the planets, since a telescope can magnify the planets to make them look like discs, but cannot do the same with the fixed stars.

Galileo related how he "was overwhelmed by the vast quantity of stars," so many that he found "more than five hundred new stars distributed among the old ones within limits of one or two degrees of arc." To three previously known stars in Orion's Belt and six in the Sword (Fig. 17), he added "eighty adjacent stars." In several pictures he presented the results of his observations with a large number of newly discovered stars amongst the older ones. Although Galileo does not make the point explicitly, it is implied that one hardly needed to put one's faith in the ancients, since they had never seen most of the stars, and had spoken from woefully incomplete evidence. A weakness of naked-eye observation was exposed by Galileo in terms of "the nature and the material of the Milky Way." With the aid of the telescope, he wrote, the Milky Way has been "scrutinized so directly and with such ocular certainty that all the disputes which have vexed philosophers through so many ages have been resolved, and we are at last freed from wordy debates about it." Seen through the telescope, the Milky Way is "nothing but a congeries of innumerable stars grouped together in clusters. Upon whatever part of it the telescope is directed, a vast crowd of stars is immediately presented to view." And this was true not only of the Milky Way, but also of "the stars which have been called 'nebulous' by every astronomer up to this time," and which "turn out to be groups of very small stars arranged in a wonderful manner." Now for the big news:

> We have . . . briefly recounted the observations made thus far with regard to the moon, the fixed stars and the Milky Way. There remains the matter which in my opinion deserves to be considered the most important of all—the disclosure of four PLANETS never seen from the creation of the world up to our own time, together with the occasion of my having discovered and studied them, their arrangements, and the observations made of their movements and alterations during the past two months. I invite all astronomers to apply themselves to examine them and determine their periodic times, something which has so far been quite impossible to complete, owing to the shortness of the time. Once more, however, warning is given that it will be necessary to have a very accurate telescope such as we have described at the beginning of this discourse.

It is interesting to observe that Galileo called the newly discovered objects "Medicean stars," although we would call them

Fig. 17. Orion's Belt and Sword, viewed through Galileo's telescope, was seen to contain eighty more stars (the smaller ones) than could be discerned by the naked eye.

PLATE I. "Will it fall back down again?" This old woodcut, taken from the correspondence of René Descartes, illustrates an experiment proposed by Father Mersenne, contemporary and friend of Galileo, to test the behavior of falling bodies. "*Retombera-t-il?*" the legend asks. Will the cannon ball come back down again?

PLATE II. A landscape like the earth's but a dead one was what impressed Galileo the first time he turned his telescope to the moon.

PLATE III. Galileo was the first to see the craters on the moon. His observations killed the ancient belief that the moon was smooth and perfectly spherical.

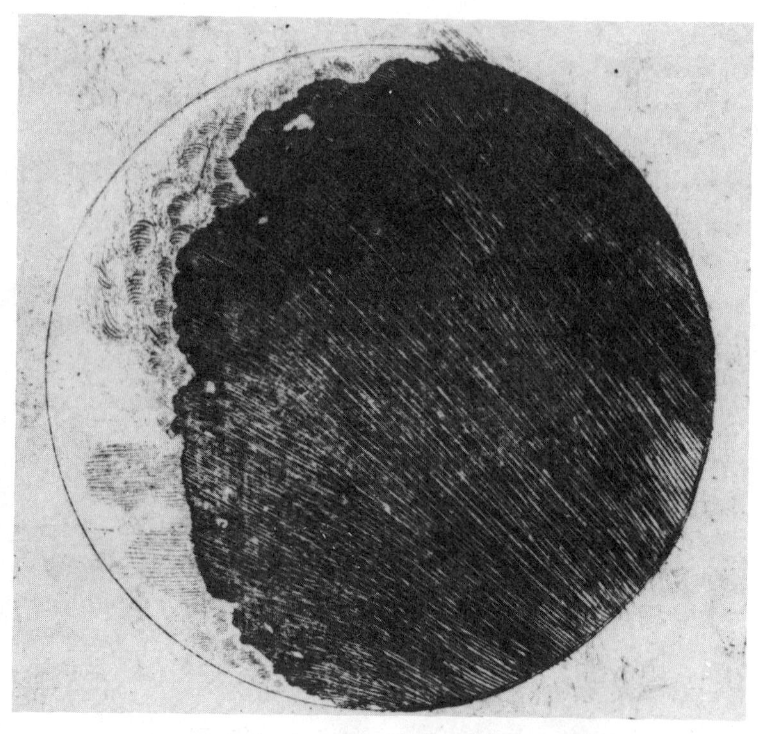

PLATE IV. Galileo's own drawing of the moon is reproduced here but upside down in accordance with the practice of showing astronomical photographs. Telescopic cameras take an inverted picture.

moons or satellites of Jupiter.* We must remember that in Galileo's day almost all the heavenly objects were called stars—a term which could include both the fixed stars and the wandering stars (or planets). Hence the newly discovered objects, which were "wanderers," and so a kind of planet, could also be called stars. Most of Galileo's book is, in fact, devoted to his methodical observations of Jupiter and the "stars" near it. Sometimes they were seen to the east and sometimes to the west of Jupiter, but never very far from the planet. They accompanied Jupiter "in both its retrograde and direct movements in a constant manner," so that it was evident that they were somehow connected to Jupiter.

JUPITER AS EVIDENCE

The first thoughts, that these might have been simply some new stars near which Jupiter was seen, were dispelled as Galileo observed that these newly discovered objects continued to move along with Jupiter. (See Supplement 2.) It was also possible for Galileo to show that the sizes of their respective orbits about Jupiter were different, and that the periodic times were likewise different. Let us allow Galileo to set forth the conclusions he drew from these observations in his own words:

> Here we have a fine and elegant argument for quieting the doubts of those who, while accepting with tranquil mind the revolutions of the planets about the sun in the Copernican system, are mightily disturbed to have the moon alone revolve about the earth and accompany it in an annual rotation about the sun. Some have believed that this structure of the universe should be rejected as impossible. But now we have not just one planet rotating about another while both run through a great orbit around the sun; our own eyes show us four stars which wander around Jupiter as does the moon around the earth, while all together trace out a grand revolution about the sun in the space of twelve years.

Jupiter, a small-scale model of the whole Copernican system, in which four small objects move around the planet just as the

*Our term "satellite" became part of the standard language of science only after it was used in this sense by Newton in his *Principia* (1687).

planets move around the bright sun, thus answered one of the major objections to the Copernican system. Galileo could not at this point explain why it was that Jupiter could move in its orbit without losing its four encircling attendants, any more than he was ever really able to explain how the earth could move through space and not lose its one encircling moon. But whether or not he knew the reason, it was perfectly plain that in every system of the world that had ever been conceived Jupiter was considered to move in an orbit, and if it could do so and not lose four of its moons, why could not the earth move without losing a single moon? Furthermore, if Jupiter has four moons, the earth can no longer be considered unique in the sense of being the only object in the universe with a moon. Furthermore, having four moons is certainly more impressive than having only one.

Although Galileo's book ends with the description of the satellites of Jupiter, it will be wise, before we explore the implications of his research, to discuss three other astronomical discoveries made by Galileo with his telescope. The first was the discovery that Venus exhibits phases. For a number of reasons Galileo was overjoyed to discover that Venus exhibits phases. In the first place, it proved that Venus shines by reflected light, and not by a light of its own; this meant that Venus is like the moon in this regard, and also like the earth (which Galileo had previously found to shine by reflected light of the sun). Here was another point of similarity between the planets and the earth, another weakening of the ancient philosophical barrier between earth and "heavenly" objects. Furthermore, as may be seen in Fig. 18A, if Venus moves in an orbit around the sun, not only will Venus go through a complete cycle of phases, but under constant magnification the different phases will appear to be of different sizes because of the change in the distance of Venus from the earth. For instance, when Venus is at such a position as to enable us to see a complete circle or almost a complete circle, corresponding to a full moon, the planet is on the opposite side of its orbit around the sun from the earth, or is seen at its farthest distance from the earth. When Venus exhibits a half circle, corresponding to a quarter moon, the planet is not so far from the earth. Finally, when we barely see a faint crescent, Venus must be at its nearest

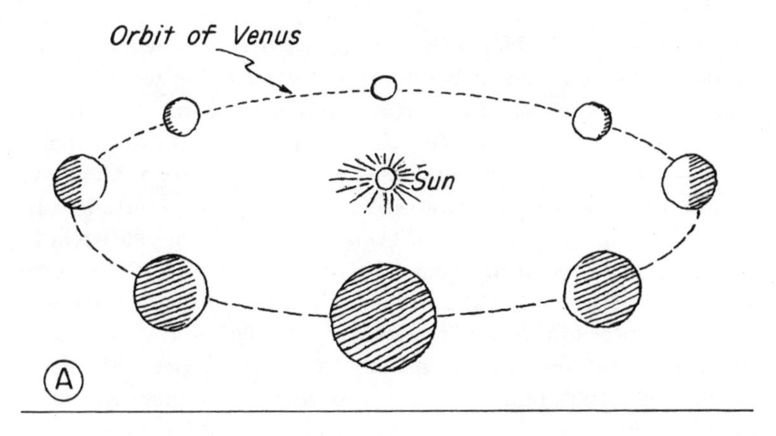

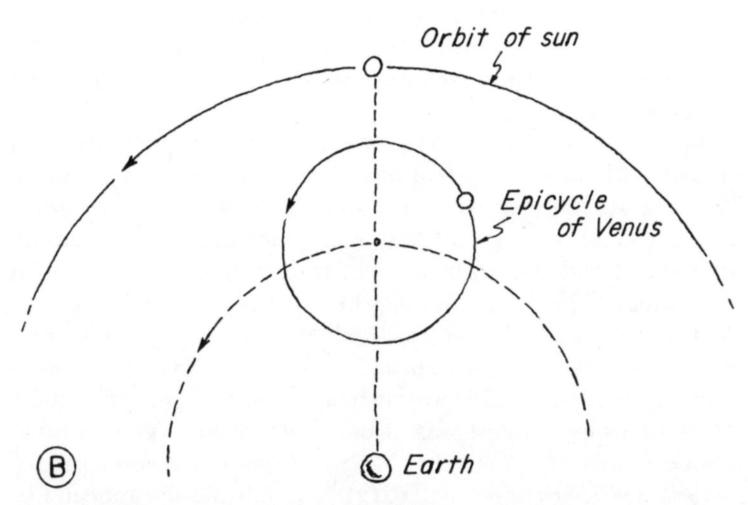

Fig. 18. The phases of Venus, first observed by Galileo, were a powerful argument against the ancient astronomy. In (A) you can see how the existence of phases accords with the system of Copernicus and how the change in the relative apparent diameter of Venus supports the concept of the planet having a solar orbit. In (B) you can see why the phenomenon would be impossible in the Ptolemaic system.

point to the earth. Hence, we should expect that when Venus shows a faint crescent it would appear very large; when Venus shows the appearance of a quarter moon, it would be of moderate size; when we see the whole disc, Venus should be very small.

According to the Ptolemaic system, Venus (like Mercury) would never be seen far from the sun, and hence would be observed only as morning star or evening star near the place where the sun has either risen or set. The center of the orbit's epicycle would be permanently aligned between the center of the earth and the center of the sun and would move around the earth with a period of one year, just as the sun does. But it is perfectly plain, as may be seen in Fig. 18B, that in these circumstances we could never see the complete sequence of phases Galileo observed—and we can observe. For instance, the possibility of seeing Venus as a disc arises only if Venus is farther from the earth than the sun; this can never occur according to the principles of the Ptolemaic system. Here then was a most decisive blow against the Ptolemaic system.

We need not say much about two further telescopic discoveries of Galileo, because they had less force than the previous ones. The first was the discovery that sometimes Saturn appeared to have a pair of "ears," and that sometimes the "ears" changed their shape and even disappeared. Galileo never could explain this strange appearance, because his telescope could not resolve the rings of Saturn. But at least he had evidence to demonstrate how erroneous it was to speak of planets as perfect celestial objects, when they could have such queer shapes. One of his most interesting observations was of the spots on the sun, described in a book that bore the title *History and Demonstrations Concerning Sunspots and Their Phenomena* (1613). Not only did the appearance of these spots prove that even the sun was not the perfect celestial object described by the ancients, but Galileo was able to show that from observations of these spots one could prove the rotation of the sun, and even compute the speed with which the sun rotates upon its axis. Although the fact that the sun does rotate became extremely important in Galileo's own mechanics, it did not imply that there must be an annual revolution of the earth around the sun.

A NEW WORLD

As may be imagined, the excitement caused by these new discoveries was communicated from person to person, and the fame of Galileo spread. Naming the satellites of Jupiter "Medicean stars" had the desired effect of obtaining for Galileo the post of mathematician to Grand Duke Cosimo of the House of Medici and enabling him to return to his beloved Florence. The discovery of the new planets was hailed as the discovery of a new world, and Galileo acclaimed the equal of Columbus. Not only did scientists and philosophers become excited by the new discoveries, but all men of learning and wit, poets and courtiers and painters, responded in the same way. A painting by the artist Cigoli for a chapel in Rome used Galileo's telescopic discoveries concerning the moon for a motif. In a poem by Johannes Faber, Galileo receives the following praise:

> *Yield, Vespucci, and let Columbus yield. Each of these*
> *Attempts, it is true, a journey through the unknown sea. . . .*
> *But you, Galileo, alone gave to the human race the sequence of stars,*
> *New constellations of heaven.*

One poem in praise of Galileo's discoveries was written by Maffeo Cardinal Barberini, who later—as Pope Urban VIII—directed that Galileo be brought to trial by the Inquisition; he told Galileo that he wanted to add lustre to his poetry by coupling it with Galileo's name. Ben Jonson wrote a masque that alludes to Galileo's astronomical discoveries; Jonson called his work *Newes from the New World*—not the new world of America but the moon, from which news can be brought through the telescope (although here it is brought by poetry). To gain some idea of the way in which this news was spread, read the following extract from a letter written on the day that Galileo's *Sidereus nuncius* appeared in Venice, March 13, 1610, by Sir Henry Wotton, the British Ambassador to Venice:

> Now touching the occurrents of the present, I send herewith unto His Majesty the strangest piece of news (as I may justly call it) that he hath ever yet received from any part of the world; which is the annexed

> book (come abroad this very day) of the Mathematical Professor at Padua, who by the help of an optical instrument (which both enlargeth and approximateth the object) invented first in Flanders, and bettered by himself, hath discovered four new planets rolling about the sphere of Jupiter, besides many other unknown fixed stars; likewise, the true cause of the *Via Lactea* [Milky Way], so long searched; and lastly, that the moon is not spherical, but endued with many prominences, and, which is of all the strangest, illuminated with the solar light by reflection from the body of the earth, as he seemeth to say. So as upon the whole subject he hath first overthrown all former astronomy—for we must have a new sphere to save the appearances—and next all astrology. For the virtue of these new planets must needs vary the judicial part, and why may there not yet be more? These things I have been bold thus to discourse unto your Lordship, whereof here all corners are full. And the author runneth a fortune to be either exceeding famous or exceeding ridiculous. By the next ship your Lordship shall receive from me one of the above instruments, as it is bettered by this man.

When Kepler wrote of Galileo's discoveries in the preface to his *Dioptrics,* he sounded more like a poet than a scientist: "What now, dear reader, shall we make of our telescope? Shall we make a Mercury's magic-wand to cross the liquid aether with, and like Lucian, lead a colony to the uninhabited evening star, allured by the sweetness of the place? Or shall we make it a Cupid's arrow, which, entering by our eyes, has pierced our inmost mind, and fired us with a love of Venus?" Enraptured, Kepler wrote, "O telescope, instrument of much knowledge, more precious than any scepter! Is not he who holds thee in his hand made king and lord of the works of God?"

In 1615, James Stephens could call his mistress "my perspective glasse, through which I view the world's vanity." And Andrew Marvell wrote of Galileo's discovery of sun spots:

> *So his bold Tube, Man, to the Sun apply'd,*
> *And Spots unknown to the bright Stars descry'd;*
> *Show'd they obscure him, while too near they please,*
> *And seem his Courtiers, are but his disease.*
> *Through Optick Trunk the Planet seem'd to hear,*
> *And hurls them off, e're since, in his Career.*

John Milton was well aware of Galileo's discoveries. Milton, whose views on the epicycle were quoted in Chapter 3, stated that when he was in Italy he "found and visited the famous Galileo, grown old a prisoner to the Inquisition." In his *Paradise Lost,* he refers more than once to the "glass of Galileo," or the "optic glass" of the "Tuscan artist," and to the discoveries made with that instrument. Writing of the moon in terms of the major phenomena discovered by Galileo, Milton referred to "new lands, rivers or mountains in her spotty globe"; and the discovery of the planets of Jupiter suggested that other planets might have their attendants too: ". . . and other Suns, perhaps with their attendant Moons, thou wilt descry." But, apart from specific references to Galileo's astronomical discoveries, what chiefly impressed Milton was the vastness of the universe and the innumerable stars described by Galileo:

. . . stars
Numerous, and every star perhaps a world
Of destined habitation.

This conveys the frightening thought of the immensity of space, and the fact that the moving earth must be a tiny pinpoint in this space with no fixed place.

Within a few years of the publication of Galileo's book, a sensitive reaction to it appeared in the works of the poet John Donne. Galileo's researches and discoveries crop up again and again in Donne's writings, and in particular *The Sidereal Messenger* is the subject of discussion in a work called *Ignatius His Conclave,* in which Galileo is described as he "who of late hath summoned the other worlds, the Stars, to come nearer to him, and give him an account of themselves." Later Donne refers to "Galilæo, the Florentine . . . who by this time hath thoroughly instructed himselfe of all the hills, woods, and Cities in the new world, the Moone. And since he effected so much with his first Glasses, that he saw the Moone, in so neere a distance that hee gave himselfe satisfaction of all, and the least parts in her, when now being growne to more perfection in his Art, he shall have made new

Glasses, . . . he may draw the Moone, like a boate floating upon the water, as neere the earth as he will."

Prior to 1609 the Copernican system had seemed a mere mathematical speculation, a proposal made to "save the phenomena." The basic supposition that the earth was "merely another planet" had been so contrary to all the dictates of experience, of philosophy, of theology, and of common sense that very few men had faced up to the awesome consequences of the heliostatic system. But after 1609, when men discovered through Galileo's eyes what the universe was like, they had to accept the fact that the telescope showed the world to be non-Ptolemaic and non-Aristotelian, in that the uniqueness attributed to the earth (and the physics based on that supposed uniqueness) could not fit the facts. There were only two possibilities open: One was to refuse to look through the telescope or to refuse to accept what one saw when one did; the other was to reject the physics of Aristotle and the old geocentric astronomy of Ptolemy.

In this book we are more concerned with the rejection of the Aristotelian physics than we are with the rejection of the Ptolemaic astronomy, except that one went with the other. Aristotelian physics, as we have seen, was based on two postulates that could not stand the Copernican assault: One was the immobility of the earth; the other was the distinction between the physics of the earthly four elements and the physics of the fifth celestial element. So we may understand that after 1610 it became increasingly clear that the old physics had to be abandoned, and a new physics established—a physics suitable for the moving earth required in the Copernican system.*

*Galileo's observations of the phases and relative sizes of Venus, and of the occasional gibbous phase of Mars, proved that Venus and presumably the other planets move in orbits around the sun. There is no planetary observation by which we on earth can prove that the earth is moving in an orbit around the sun. Thus all Galileo's discoveries with the telescope can be accommodated to the system invented by Tycho Brahe just before Galileo began his observations of the heavens. In this Tychonic system, the planets Mercury, Venus, Mars, Jupiter, and Saturn move in orbits around the sun, while the sun moves in an orbit around the earth in a year. Furthermore, the daily rotation of the heavens is communicated to the sun and planets, so that the earth itself neither rotates nor revolves in an orbit. The Tychonic system appealed to those who sought to save the immobility of the earth while accepting some of the Copernican innovations.

But for most thinkers in the decades following Galileo's observations with the telescope, the concern was not so much for the need of a new system of physics, as it was for a new system of the world. Gone forever was the concept that the earth has a fixed spot in the center of the universe, for it was now conceived to be in motion, never in the same place for any two immediately successive instants. Gone also was the comforting thought that the earth is unique, that it is an individual object without any likeness anywhere in the universe, that our uniqueness requires a unique habitation. There were other problems that soon arose, of which one is the size of the universe. For the ancients the universe was finite, each of the celestial spheres, including that of the fixed stars, being of finite size and moving in its diurnal motion so that each part of it had a finite speed. If the stars were at an infinite distance, then they could not move in a daily circular motion around the earth with a finite speed, for the path of an object at an infinite distance must be infinitely long, and the time it takes to move an infinite distance cannot be finite. Hence in the geostatic system the fixed stars could not be infinitely far away. But in the Copernican system, when the fixed stars were not only fixed with regard to one another but were actually considered fixed in space, there was no such limitation upon their distance.

Not all Copernicans considered the universe infinite, and Copernicus himself certainly thought of the universe as finite, as did Galileo. But others saw Galileo's discoveries as indicating the presence of innumerable stars at infinite distances, and the earth itself diminished to a speck. The image of the disruption of "this little world of man," and what has been called "the realization how slight a part that world plays in an enlarged and enlarging universe," was brilliantly expressed in these lines of a sensitive clergyman and poet, John Donne:

And new Philosophy calls all in doubt,
The Element of fire is quite put out;
The Sun is lost, and th' earth, and no mans wit
Can well direct him where to looke for it.

And freely men confesse that this world's spent,
When in the Planets, and the Firmament
They seeke so many new; then see that this
Is crumbled out againe to his Atomies.
'Tis all in peeces, all cohaerence gone;
All just supply, and all Relation.

CHAPTER
5

Toward an Inertial Physics

After the second decade of the seventeenth century, the reality of the Copernican system was no longer an idle speculation. Copernicus himself, understanding the nature of his arguments, had stated quite explicitly, in the preface to *On the Revolutions of the Celestial Spheres,* that "mathematics is for the mathematicians." Another preface, unsigned, emphasized the disavowal. Inserted in the book by Osiander, a German clergyman into whose hands the printing had been entrusted, the second preface said that the Copernican system was not presented for debate on its truth or falsity, but was merely another computing device. This was all very well until Galileo made his discoveries with the telescope; then it became urgent to solve the problems of the physics of an earth in motion. Galileo devoted a considerable portion of his intellectual energy toward this end, and with a fruitful result, for he laid the foundations of the modern science of motion. He tried to solve two separate problems: first, to account for the behavior of falling bodies on a moving earth, falling exactly as they would appear to do if the earth were at rest, and, second, to establish new principles for the motion of falling bodies in general.

UNIFORM LINEAR MOTION

Let us begin by a consideration of a limited problem: that of uniform linear motion. By this is meant motion proceeding in a straight line in such a way that if any two equal intervals of time are chosen, the distance covered in those two intervals will always be identical. This is the definition Galileo gave in his last and perhaps greatest book, *Discourses and Demonstrations Concerning Two New Sciences,* published in 1638, after his trial and condemnation by the Roman Inquisition.* In this book Galileo presented his most mature views on the new science of motion he had founded. He emphasized particularly the fact that in defining uniform motion, it is important to make sure that the word "any" is included, for otherwise, he said, the definition would be meaningless. In this he was certainly criticizing some of his contemporaries and predecessors.

Suppose that there is such a motion in nature; we may ask with Galileo, what experiments could we imagine to demonstrate its nature? If we are in a ship or carriage moving uniformly in a straight line, what actually will happen to a weight allowed to fall freely? The answer, experiment will prove, is that in such circumstances the falling will be straight downward with regard to the frame of reference (say the cabin of a ship, or the interior of a carriage), and it will be so whether that frame of reference is standing still with regard to the outside environment or moving forward in a straight line at constant speed. Expressing it differently, we may state the general conclusion that no experiment can be performed within a sealed room moving in a straight line at constant speed that will tell you whether you are standing still or moving. In actual experience, we can often tell whether we are standing still or moving, because we can see from a window whether there is any relative motion between us and the earth. If the room is not closely sealed, we may feel the air rushing through and creating a wind. Or we may feel the vibration of motion or hear the wheels turning in a carriage, automobile, or railroad car. A form of relativity is involved here, and it was stated very clearly by Copernicus, because it was essential to his argument to establish that when two objects, such as the sun and earth, move relative to each other, it is impossible to tell which one is at rest and which one is in motion. Copernicus could point to the example of two ships at harbor, one pulling away from the

*This work was published in Leyden. Galileo evidently did not approve of the title (given to the book by the publisher), which "he considered to be undignified and ordinary."

other. A man on a ship asks which of the two, if either, is at anchor and which is moving out with the tide. The only way to tell is to observe the land, or a third ship at anchor. In present-day terms, we could use for this example two railroad trains on parallel tracks facing in opposite directions. Many of us have had the experience of watching a train on the adjacent track and thinking that we are in motion, only to find when the other train has left the station that we have been at rest all the time.

A LOCOMOTIVE'S SMOKESTACK AND A MOVING SHIP

But before we discuss this point further, an experiment is in order. This demonstration makes use of a toy train traveling along a straight track with what closely approximates uniform motion. The locomotive's smokestack contains a small cannon actuated by a spring, so constructed that it can fire a steel ball or marble vertically into the air. When the gun is loaded and the spring set, a release underneath the locomotive actuates a small trigger. In the first part of this experiment the train remains in place upon the track. The spring is set, the ball placed in the small cannon, and the release mechanism triggered. In Plate 6A, a scene of successive stroboscopic photos shows the position of the ball at equally separated intervals. Observe that the ball travels straight upward, reaches its maximum, then falls straight downward onto the locomotive, thus striking almost the very point from which it had been shot. In the second experiment the train is set into uniform motion, and the spring once again released. Plate 6B shows what happens. A comparison of the two pictures will convince you, incidentally, that the upward and downward part of the motion is the same in both cases, and is independent of whether the locomotive is at rest or has a forward motion. We shall come back to this later in the chapter, but for the present we are primarily concerned with the fact that the ball continued to move in a forward direction with the train, and that it fell onto the locomotive just as it did when the train was at rest. Plainly then, this particular experiment, at least to the extent of determining whether the ball returns to the cannon or not, will never tell us whether the train is standing still or moving in a straight line with a constant speed.

Even those who cannot explain this experiment can draw a most important conclusion. Galileo's inability to explain how Jupiter could move without losing its satellites did not destroy the phenomenon's effectiveness as an answer to those who asked how the earth could move and not lose its moon. Just so our train experiment—even if unexplainable—would be sufficient answer to the argument that the earth must be at rest because otherwise a dropped ball would not fall vertically downward to strike the ground at a point directly below, and a cannon ball shot vertically upward would never return to the cannon.

It should be observed, and this is an important point to which we shall return in a later chapter, that the experiment we have just described is not exactly related to the true situation of a moving earth, because in the earth's daily rotation each point on its surface is moving in a circle while in its annual orbit the earth is traveling along a gigantic ellipse. It is nevertheless true that for ordinary experiments, in which the falling motion would usually occupy only a few seconds, or at most a few minutes, the departure of the motion of any point on the earth from a straight line is small enough to be insignificant.

Galileo would have nodded in approval at our experiment. In his day the experiment was discussed, but not often performed. (For Galileo's inertial experiments, see Supplement 9.) The usual reference frame was a moving ship. This was a traditional problem, which Galileo introduced in his famous *Dialogue Concerning the Two Chief World Systems,* as a means of confuting the Aristotelian beliefs. In the course of this discussion, Galileo has Simplicio, the character in the dialogue who stands for the traditional Aristotelian, say that in his opinion an object dropped from the mast of a moving ship will strike the ship somewhere behind the mast along the deck. On first questioning, Simplicio admits that he has never performed the experiment, but he is persuaded to say that he assumes that Aristotle or one of the Aristotelians must have done this experiment or it would not have been reported. Ah no, says Galileo, this is certainly a false assumption, because it is plain that they have never performed this experiment. How can Galileo be so sure? asks Simplicio, and he receives this reply: The proof that this experiment was never performed lies in the fact that the wrong answer was obtained. Galileo has given the right answer. The object will fall at the foot of the mast, and it will do so whether the ship is in motion or whether the ship is at rest. Incidentally, Galileo asserted elsewhere that he had performed such an experiment, although he did not say so in his treatise. Instead he said, "I, without experimenting, know that the result must be as I say, because it is necessary."

Why is it that an object falls to the same spot on the deck from the mast of a ship that is at rest and from the mast of a ship that is moving in a straight line with constant speed? For Galileo it was not enough that this should be so; it required some principle that would be basic to a system of physics that could account for the phenomena observed on a moving earth.

GALILEO'S SCIENCE OF MOTION

Our toy train experiment, to which we shall refer again in the last chapter, illustrates three major aspects of Galileo's work on motion. In the first place, there is the principle of inertia, toward which Galileo strove but which, as we shall see in the final chapter, awaited the genius of Isaac Newton for its modern definitive formulation. Secondly, the photographs of the distances of descent of the ball after successive equal intervals of time illustrate his principles of uniformly accelerated motion. Finally, in the fact that the rate of downward fall during the forward motion is the same as the rate of downward fall at rest, we may see an example of Galileo's famous principles of the independence and composition of vector velocities.

We shall examine these three topics by first considering Galileo's studies of accelerated motion in general, then his work dealing with inertia, and finally his analysis of complex motions.

In studying the problem of falling bodies, Galileo, we know, made experiments in which he dropped objects from heights, and—notably in the Pisan days of his youth—from a tower. Whether the tower was the famous Leaning Tower of Pisa or some other tower we cannot say; the records that he kept merely tell us that it was from some tower or other. Later on his biographer Viviani, who knew Galileo during his last years, told a fascinating story that has since taken root in the Galileo legend. According to Viviani, Galileo, desiring to confute Aristotle, ascended the Leaning Tower of Pisa, "in the presence of all other teachers and philosophers and of all students," and "by repeated experiments" proved "that the velocity of moving bodies of the same composition, unequal in weight, moving through the same medium, do not attain the proportion of their weight, as Aristotle assigned it to them, but rather that they move with equal velocity. . . ." Since there is no record of this public demonstration in any other source, scholars have tended to doubt that it happened, especially since in its usual telling and retelling it becomes fancier each time. Whether Viviani made it up, or whether Galileo told it to him in his old age, not remembering exactly what had happened many decades earlier, we do not know. But the fact of the matter is that the results do not agree with those given by Galileo himself because, as we mentioned in an earlier chapter, Galileo pointed out very carefully that bodies of unequal weight do not attain quite the same velocity, the heavier member of the pair striking the ground a little before the lighter.

Such an experiment, if performed, could only have the result of proving Aristotle wrong. In Galileo's day, it was hardly a great achievement to prove that Aristotle was wrong in only one respect. Pierre de la Ramée (or Ramus) had some decades earlier made it known that everything in Aristotle's physics was unscientific. The inadequacies of the Aristotelian law of motion had been evident for at least four centuries, and during that time a considerable body of criticism had piled up. Although they struck another blow at Aristotle, experiments from the tower, whether the Tower of Pisa or any other, certainly did not disclose to Galileo a new and correct law of falling bodies. Yet formulation of the law was one of his greatest achievements. (See Supplement 4.)

To appreciate the full nature of Galileo's discoveries, we must understand the importance of abstract thinking, of its use by Galileo as a tool that in its ultimate polish was a much more revolutionary instrument for science than even the telescope.

Galileo showed how abstraction may be related to the world of experience, how from thinking about "the nature of things," one may derive laws related to direct observation. In this process, experiment was of paramount importance to Galileo, as we have recently learned, thanks largely to the ingenious researches of Stillman Drake. Let us now outline the main stages of Galileo's thought processes, as he described them to us in his *Two New Sciences.*

Galileo says:

> There is perhaps nothing in nature older than motion, about which volumes neither few nor small have been written by philosophers; yet I find many essentials . . . of it that are worth knowing which have not even been remarked, let alone demonstrated.

Galileo recognized that others before him had observed that the natural motion of a heavy falling body is continuously accelerated. But he said that it was his achievement to find out "the proportion according to which this acceleration takes place." He was proud that it was he who had found for the first time "that the spaces run through in equal times by a moveable descending from rest maintain among themselves the same rule . . . as do the odd numbers following upon unity." He also proved that "missiles or projectiles" do not merely describe a curved path of some sort; the path in fact is a parabola.

In discussing Galileo's thoughts on motion, we have two very different options. One is to try to trace the development of his ideas through his manuscripts and correspondence and other documents, the other to summarize the public presentation that he published in his *Discourses and Demonstrations Concerning Two New Sciences.* The first of these is necessarily tentative, since in part it depends on the interpretation of certain manuscript pages containing numerical data and diagrams without any commentary or explanation (see Supplement 4); this is the private record, of which the decipherment began only in the 1970s. The second option, the public record, comprises the presentation that Galileo intended to have us study. It is this public (published) presentation that has actually conditioned the advance of science, in the domain of motion, from Galileo's revolutionary new kinematics to the modern science of dynamics. We call Galileo's subject kinematics because it was largely a study of uniform and accelerated motions without much consideration of the forces, whereas dynamics discloses the forces acting on bodies to produce or to alter motion, and the laws relating the forces to the changes in motion they produce. Although Galileo was aware that accelerations result from the actions of forces (e.g., the acceleration of falling being produced by the force of bodies' weights), he did not concentrate on this part of the topic. Yet, because Galileo did give consideration to forces and motions in some special but important cases, we should perhaps describe his subject as kinematics with some dynamics. Newton believed that Galileo had known and made use of the first two of his own three "axioms or laws of motion," embodying the most fundamental principles of dynamics.

First, Galileo discusses the laws of uniform motion, in which the distance is proportional to the time and the speed, therefore, constant. Then he turns to accelerated motion. To "seek out and clarify the definition that best agrees with that [accelerated motion] which nature employs" is for Galileo the primary problem. Anyone may invent "at pleasure some kind of motion," he says. But, "since nature does employ a certain kind of acceleration for descending heavy things," he "decided to look into their properties" in order to be sure that the definition of accelerated motion that he was about to use would agree "with the essence of naturally accelerated motion." Galileo says, furthermore, that "in the investigation of naturally accelerated motion," we shall be led, "by the hand," as it were, "by consideration of the custom and procedure of nature herself, in all her other works," in "the performance of which she habitually employs the first, simplest, and easiest means." Galileo was invoking a famous principle here, one that actually goes back to Aristotle, that nature always works in the simplest way possible, or in the most economical fashion. Galileo says:

> When . . . I consider that a stone, falling from rest, successively acquires new increments of speed, why should I not believe that those additions are made by the simplest and most evident rule? For if we look into this attentively, we can discover no simpler addition and increase than that which is added on always in the same way.

Proceeding on the principle that nature is simple, so that the simplest change is one in which the change itself is constant, Galileo states that if there is an equal increment of speed in each successive interval of time, this is plainly the simplest possible accelerated motion. Shortly thereafter, Galileo has Simplicio (the Aristotelian) say that he holds to a different belief, namely, that a falling body has a "velocity increasing in proportion to the space," and we as critical readers must admit it certainly seems to be as "simple" as Galileo's definition of accelerated motion. Of the two possibilities

$$V \propto T \qquad [1]$$

$$V \propto D \qquad [2]$$

which is simpler? Are not both examples of "an increment . . . which repeats itself always in the same manner," either the same increment in speed in equal time intervals or the same increment in equal spaces? They are equally simple because both are equations of the first degree, both examples of a simple proportionality. Both are therefore much simpler than any of the six possibilities that follow:

$$V \propto \frac{1}{T} \qquad [3]$$

$$V \propto \frac{1}{T^2} \qquad [4]$$

$$V \propto T^2 \qquad [5]$$

$$V \propto \frac{1}{D} \qquad [6]$$

$$V \propto \frac{1}{D^2} \qquad [7]$$

$$V \propto D^2 \qquad [8]$$

On what possible ground can we reject the relationship suggested by Simplicio and given in Equation (2)? Since each of Equations (1) and (2) is formally as simple as the other, Galileo is forced to introduce another criterion for his choice. He asserts that possibility No. 2—the speed increases in proportion to the distance fallen—will lead to a logical inconsistency, while the relationship given in Equation (1) does not. Hence it would appear that since one of the "simple" assumptions leads to an inconsistency, while the other does not, the only possibility is that falling bodies have speeds that increase in proportion to the time in which they have fallen.

This conclusion, as presented in Galileo's last and most mature work, has a special interest for the historian, because the argument whereby Galileo "proves" that a logical inconsistency follows from Equation (2) contains an error. There is no "logical" inconsistency here; the problem is merely that this relation is incompatible with the assumption of a body starting from rest. The historian is also interested to discover that earlier in his life Galileo wrote about this very same subject in a wholly different way to his friend Fra Paolo Sarpi. In this letter Galileo appears to have believed that the speed of freely falling bodies increases in direct proportion to the distance fallen. From this assumption, Galileo believed that he could deduce that the distance fallen must be proportional to the square of the time, or that the assumption of Equation (2) leads to the equation

$$D \propto T^2. \qquad [9]$$

Then Galileo goes on to say that the proportionality of the distance to the square of the time is "well known." Between writing the letter to Sarpi, and the appearance of the *Two New Sciences,*

Galileo had corrected this apparent error. (See Supplement 5.)

In any event, in the *Two New Sciences,* Galileo proves that the relationship shown in Equation (9) follows from Equation (1). Galileo does so by means of an ancillary theorem as follows:

> Proposition I. Theorem I. The time in which a certain space is traversed by a moveable in uniformly accelerated movement from rest is equal to the time in which the same space would be traversed by the same moveable carried in uniform motion whose degree of speed is one-half the maximum and final degree of speed of the previous, uniformly accelerated, motion.

By using this theorem, and the theorems on uniform motion, Galileo proceeds to

> Proposition II. Theorem II. If a moveable descends from rest in uniformly accelerated motion, the spaces run through in any times whatever are to each other as the duplicate ratio of their times [that is, are as the squares of those times].

This is the result expressed in Equation (9), and it leads to Corollary 1. In this corollary Galileo shows that if a body falls from rest with uniformly accelerated motion, then the spaces $D_1, D_2, D_3, \ldots$ which are traversed in successive equal intervals of time "will be to one another [in the same ratio] as are the odd numbers, starting from unity, that is, as 1, 3, 5, 7. . . ." Galileo is quick to point out that this series of odd numbers is derived from the fact that the distances gone in the first time interval, the first two time intervals, the first three time intervals, . . . are as the squares 1, 4, 9, 16, 25, . . . ; the differences between them are the odd numbers. The conclusion is of a special interest to us, because it was part of the Platonic tradition to believe that the fundamental truths of nature were disclosed in the relations of regular geometrical figures and relations between numbers, a point of view to which Galileo expresses his devotion in an earlier part of the book. He has Simplicio say: "Believe me," if "I were to begin my studies over again, I should try to follow the advice of Plato and commence from mathematics, which proceeds so carefully, and does not admit as certain anything except what it has conclusively proved." To Galileo it is evidently a token of the soundness of his discussion of falling bodies that he may conclude: "Thus when the degrees of speed are increased in equal times according to the simple series of natural numbers, the spaces run through in the same times undergo increases that accord with the series of odd numbers from unity."*

*The stages whereby Galileo proceeds (in the *Two New Sciences*) from the definition of uniformly accelerated motion

$$V \propto T$$

to the law of accelerated motion or the law of free fall (the time-squared law)

$$D \propto T^2$$

are easy to rewrite in simple algebraic language. In a time T_0, starting from rest, the body acquires a speed V_0. The average or mean speed is $\frac{1}{2} V_0$. The distance traversed under acceleration during time T_0 is the same as if the body had moved during that same time interval with a constant speed equal to the average speed. The distance D_0 at constant speed $\frac{1}{2} V_0$ is

$$D_0 = \tfrac{1}{2} V_0 T_0.$$

But since

$$V_0 \propto T_0$$

it follows that

$$D_0 = \tfrac{1}{2} V_0 T_0 \propto T_0^2.$$

To see how Galileo's numerical sequences follow from the time-squared law for distance, let the time intervals be $T, 2T, 3T, 4T, 5T, \ldots$ Then the distances will be as $T^2, 4T^2, 9T^2, 16T^2, 25T^2, \ldots$, or as 1, 4, 9, 16, 25, . . . The distances gone in the first, the second, the third, the fourth, the fifth . . . time intervals will then be as differences between successive members of this series or as 1, 3, 5, 7, 9, ... If the constant of acceleration in the uniformly accelerated motion is A, so that $V = AT$, then the last equation becomes (for D_0, V_0, T_0)

$$D_0 = \tfrac{1}{2}(V_0)T_0 = \tfrac{1}{2}(AT_0)T_0 = \tfrac{1}{2}AT_0^2$$

and in general

$$D = \tfrac{1}{2}AT^2$$

Although the numerical aspect of the investigation is satisfying to Salviati (the character in the *Two New Sciences* who speaks for Galileo), and to Sagredo (the man of general education and good will who usually supports Galileo), Galileo recognizes that this Platonic point of view can hardly satisfy an Aristotelian. Galileo therefore has Simplicio declare that he is

> able to see why the matter must proceed in this way, once the definition of uniformly accelerated motion has been postulated and accepted. But I am still doubtful whether this is the acceleration employed by nature in the motion of her falling heavy bodies. Hence, for my understanding and for that of other people like me, I think that it would be suitable at this place [for you] to adduce some experiment from those (of which you have said that there are many) that agree in various cases with the demonstrated conclusions.

Galileo agrees that Simplicio is speaking "like a true scientist" and that he has made a "very reasonable demand." There follows a description of a famous experiment. Let us allow Galileo to tell it in his own words:

> In a wooden beam or rafter about twelve *braccia* [yards] long, half a *braccio* wide, and three inches thick, a channel was rabbeted in along the narrowest dimension, a little over an inch wide and made very straight; so that this would be clean and smooth, there was glued within it a piece of vellum, as much smoothed and cleaned as possible. In this there was made to descend a very hard bronze ball, well rounded and polished, the beam having been tilted by elevating one end of it above the horizontal plane from one to two *braccia,* at will. As I said, the ball was allowed to descend along [*per*] the said groove, and we noted (in the manner I shall presently tell you) the time that it consumed in running all the way, repeating the same process many times, in order to be quite sure as to the amount of time, in which we never found a difference of even the tenth part of a pulse-beat.

---

the familiar equation for Galileo's time-squared law found in science textbooks. For the special case of free fall, the constant is denoted by g (sometimes called the "acceleration of gravity"), so that this equation becomes

$$D = \tfrac{1}{2}gT^2$$

where g has the numerical value of approximately 32 ft/sec in each second or 980 cm/sec in each second.

To this Simplicio replies: "It would have given me great satisfaction to have been present at these experiments. But being certain of your diligence in making them and your fidelity in relating them, I am content to assume them as most certain and true."

Galileo's procedure, such as we have been describing, differs radically from what is commonly described in elementary textbooks as "*the* scientific method." For in all such accounts, the first step is said to be to "collect all the relevant information," and so on. The usual method of procedure, we are told, is to collect a large number of observations, or to perform a series of experiments, then to classify the results, generalize them, search for a mathematical relation, and, finally, to find a law. But Galileo presents himself in a different mode—thinking, creating ideas, usually working with pencil or pen and paper. Of course, Galileo was not a mere "armchair" philosopher, a pure speculator. We now know that he had been making experiments and that his creative thinking was characterized by a constant interaction between abstraction and reality, between theoretical ideas and experimental data. In the *Two New Sciences,* however, Galileo stresses the bold general principle that nature is simple. He gives us an image of an experimental scientist whose thoughts are directed by abstractions of nature. He seeks for simple relationships of the first degree rather than those of a higher order. He aims to find the simplest relationship that does not lead to a contradiction. Yet, even though experiment had been a guiding force in the development of his ideas, when it came to the final presentation, the experiment of the inclined plane served as a confirming rather than an investigative experiment, and it was introduced by Galileo in response to the demand of the Aristotelian Simplicio, the spokesman for the doctrine Galileo was criticizing. Galileo presents the account of the experiment with a preliminary statement that explains carefully the purpose of such an experiment. It will be profitable for us to examine this paragraph (put by Galileo into the mouth of Salviati):

> Like a true scientist, you make a very reasonable demand, for this is usual and necessary in those sciences which apply mathematical

> demonstrations to physical conclusions, as may be seen among writers on optics, astronomers, mechanics, musicians, and others who confirm their principles with sensory experiences that are the foundations of all the resulting structure. I do not want to have it appear a waste of time [*superfluo*] on our part, [as] if we had reasoned at excessive length about this first and chief foundation upon which rests an immense framework of infinitely many conclusions—of which we have only a tiny part put down in this book by the Author, who will have gone far to open the entrance and portal that has until now been closed to speculative minds. Therefore as to the experiments: the Author has not failed to make them, and in order to be assured that the acceleration of heavy bodies falling naturally does follow the ratio expounded above, I have often made the test . . . and in his company.

It is certainly made clear by Galileo in this statement that the purpose of these experiments on an inclined plane was not to find the law in its original discovery, but rather to make certain that in fact such accelerations as Galileo discussed may actually occur in nature. It has been a cause of real astonishment to find that Galileo had in fact made his discovery of the laws of motion in a manner quite different from the public presentation he gave in his last treatise, the *Two New Sciences.* His secret was well kept for over three and a half centuries, until Stillman Drake found and drew attention to Galileo's work sheets, which seem unquestionably to be the record of experiments on moving bodies, somehow related to the laws of motion he had found. This is one of the great discoveries in the history of science of our time, even though there is not as yet a universal assent to any single interpretation of Galileo's stages of thought. (On this topic, see Supplement 4, with reference to the research of Winifred L. Wisan and R. H. Naylor; see also the article by M. Segre in the Guide to Further Reading, p. 242.) The experiment described in the *Two New Sciences,* however, is of a different kind. But observe that in point of fact what is demonstrated in such a series of experiments is not that speed is proportional to time, but only that distance is proportional to the square of the time. Since this is a result *implied* by speed's being proportional to time, it is assumed that the experiment also justifies the principle that speed is proportional to time. (See Supplement 6.)

And it is further to be noted that in introducing the experiment, Salviati says that he himself had made this particular set of observations in Galileo's company in order "to be assured that the acceleration of heavy bodies falling naturally does follow the ratio expounded above." And yet this particular set of observations of balls rolling down inclined planes does not apparently have anything to do with acceleration of freely falling bodies. In free fall, bodies have a motion that is totally unimpeded save for the small effect of air resistance. But here the body's motion is far from free, since the body is constrained to the surface of the plane. In both cases, however, the acceleration is produced by gravity. In the experiments on the inclined plane, the falling effect of gravity is "diluted," only a part of gravity acting in the direction of the plane. In these experiments it is found that distance is proportional to the square of the time at any inclination one may give the plane, however steep. The experiments are related to free fall because it may be assumed that in the limiting case, in which the plane is vertical, one can expect the law still to hold. But in that limiting case of free fall, the ball will not roll in its downward movement as it does along an inclined plane—a point that Galileo nowhere mentions. And yet this is a most important condition, because we know today from theoretical mechanics that this is a chief factor that would prevent the experiments from yielding an accurate numerical value for the acceleration of free fall. That is, one cannot use the method of components to get the acceleration of free fall from the acceleration along the inclined plane, because in one case the descent is accompanied by rolling and in the other it is not. Hence it would be far from obvious to a hard-nosed skeptic that the inclined plane experiment showed that free fall is uniformly accelerated, or even that free fall accords with the time-squared law for distance, although the experiments did show that the time-squared law occurs in nature and hence that in nature there are uniformly accelerated motions.

In our own times a number of scholars have replicated Galileo's experiment of the inclined plane; the first to do so was Thomas B. Settle. The results fully agree with Galileo's report that for various lengths,

> by experiments repeated a full hundred times, the spaces were always found to be to one another as the squares of the times. And this [held] for all inclinations of the plane; that is, of the channel in which the ball was made to descend, where we observed also that the times of descent for diverse inclinations maintained among themselves accurately that ratio that we shall find later assigned and demonstrated by our Author.

Today we find no problems in accepting Galileo's statement that "these operations repeated time and time again never differed by any notable amount" and that the accuracy of the experiment was such that the difference between two observations never exceeded "a difference of even the tenth part of a pulse-beat."

Galileo was not greatly concerned to make measurements of the times for the free vertical fall of a body. He supposed that such data could be obtained from experiments made with balls rolling down inclined planes, not appreciating the difference between freely sliding motion down the plane and rolling. In his published writings on motion, Galileo did not include any computation of the acceleration of a freely falling body by taking the limit of motion on an inclined plane. In a letter to Baliani, however, Galileo did explain a way of using inclined-plane experiments to determine the speed (and hence the acceleration) of free vertical falling motion.

In the Second Day of his *Dialogue Concerning the Two Chief World Systems,* Galileo computed the time it would take for a cannon ball to fall from the moon to the earth. In "repeated experiments," he wrote, an iron ball weighing 100 pounds "falls from a height of 100 yards in five seconds." Galileo's actual words (*Dialogue Concerning the Two Chief World Systems,* Second Day, trans. Stillman Drake, p. 223) are: ". . . let us suppose we want to make the computations for an iron ball of 100 pounds which in repeated experiments falls from a height of 100 yards in 5 seconds." Using the familiar law that $D=1/2gT^2$, Drake finds that these "repeated experiments" yield a value for the acceleration of free fall (g) of 467 cm/sec^2 as against 980 cm/sec^2. (See, further, Drake's discussion on p. 480 of his translation.) During discussions of this topic with me, Drake has informed me that "a still unpublished working paper bears Galileo's calculation of fall through 45¼ meters in 3.11 seconds, the actual time being 3.04 seconds."

Galileo himself discussed these data in his letter to Baliani of 1 Aug. 1639 (translated in Drake's *Galileo at Work*). Baliani had written in 1632 to ask Galileo how he knew that a heavy body falls through 100 yards *(braccia)* in five seconds, adding that in Genoa there was no tower of that height from which to try the experiment; he was also concerned about the distance of four yards fallen during the first second, which was extremely hard to verify. When Galileo replied some years later, he admitted that if Baliani attempted to verify by "experiment whether what I wrote about the 100 *braccia* in five seconds be true," Baliani might "find it false." He explained that the purpose of the argument was to confute Father Scheiner, who had written concerning the time for a cannon ball to fall from the moon to the earth; for Galileo's own computation of the time of fall, "it mattered little whether the five seconds for 100 *braccia* was true or not." More significant for us is Galileo's false assumption that in falling from the moon to the earth a cannon ball would keep a constant acceleration.*

*Galileo's mode of computing free fall was to deduce the value from motion on an inclined plane. As he explained to Baliani in 1639 (*Galileo at Work,* pp. 399–400): ". . . the descent of that ball that I make descend through a channel, arbitrarily sloped, will give us all the times—not only of 100 *braccia,* but of any other quantity of vertical fall—inasmuch as (as you yourself have demonstrated) the length of the said channel, or let us call it inclined plane, is a mean proportional between the vertical height of the said plane and the length of the whole vertical distance that would be passed in the same time by the falling moveable. Thus, for example, assuming that the said channel is 12 *braccia* long and its vertical height is one-half *braccio,* one *braccio,* or two, the distance passed in the vertical will be 288, 144, or 72 *braccia,* as is evident. It now remains that we find the amount of time of descent through the channel. This we shall obtain from the marvelous property of the pendulum, which is that it makes all its vibrations, large or small, in equal times." To reduce the motion of a given pendulum to seconds, Galileo further explained, it would be necessary to calibrate it by counting the number of vibrations during 24 hours, as determined by a group of "two or three or four patient and curious friends." They would mark the passage of 24 hours from the instant that a 'fixed star' "stands against some fixed marker" until "the return of the 'fixed star' to the original point." Galileo's letter to Baliani suggests this as a method for finding the distance fallen in some given time, but does not explicitly declare that he himself has performed these quantitative experiments. This might argue that, contrary to the apparent sense of Galileo's *Dialogue* (with the phrase "repeated experiments"), as interpreted by Mersenne and others, Galileo was only introducing numbers for the sake of argument.

Galileo's actual sentence in the *Dialogue* reads as if "in repeated experiments" the iron ball of 100 pounds had been observed to fall from a height of 100 yards in 5 seconds. Could it be, however, that Galileo was only supposing that this result could be obtained "in repeated experiments"? Was this what Galileo meant, that he would only suppose we want to make a calculation? If he was merely writing *ex suppositione,* then he would have been saying, in effect, "let us *assume* that experience showed a fall of 100 *braccia* to take 5 seconds," not that "repeated tests have shown this." His sentence is syntactically ambiguous.

But at least one of Galileo's contemporaries, Father Marin Mersenne, read the text in an obvious way and concluded that Galileo had alleged he had found the result he gave by "repeated experiments." Galileo "supposes," Mersenne wrote to Nicolas Claude Fabri de Peiresc on 15 Jan. 1635, "that a bullet [*boulet*] falls one hundred *braccia* in 5 seconds; wherefrom it follows that the bullet will fall not more than four *braccia* in one second." Mersenne himself was convinced that "it will fall [in one second] from a greater height." In his *Harmonie universelle* (Paris, 1636, vol. 1, p. 86), Mersenne wrote at length concerning the difference between the numerical results he obtained in Paris and its environs and those Galileo reported from Italy. He regretted that he might seem to be reproaching "such a great man for [having taken] little care in his experiments." It is still a puzzle why a careful experimenter like Galileo should have given such a poor value. Perhaps he was suggesting a "round number" for easy calculation, but in that case why write "in repeated experiments"?

In retrospect, it is clear to us that in Galileo's presentation in the *Two New Sciences,* the experiment of the inclined plane was introduced to serve as a check to see whether the principles that he had derived by the method of abstraction and mathematics actually applied in the world of nature. So far as the prospective reader is concerned, the truth of Galileo's law of falling bodies was guaranteed in the first instance by the correctness of the logic and the definitions, by the exemplification of the simplicity of nature and the relations of integers, and not merely by a series of experiments or observations. Galileo was here possibly displaying the same attitude as in his discussion of the falling of an object from the mast of a ship, where again it was the nature of things and necessary relations that counted, rather than particular sets of experiences. The correct result is to be maintained, according to Galileo, even in the face of evidence from the senses (in a form of experiments or observations) which may be antagonistic. Nowhere did Galileo express this point of view more strongly than in discussing the evidence of the senses against the motion of the earth. "For the arguments against the whirling of the earth which we have already examined are very plausible, as we have seen," Galileo wrote, "and the fact that the Ptolemaics and Aristotelians and all their disciples took them to be conclusive is indeed a strong argument of their effectiveness. But the experiences which overtly contradict the annual movement are indeed so much greater in their apparent force that, I repeat, there is no limit to my astonishment when I reflect that Aristarchus and Copernicus were able to make reason so conquer sense that, in defiance of the latter, the former became mistress of their belief" (*Dialogue Concerning the Two Chief World Systems*).

To recapitulate, Galileo demonstrated mathematically that a motion starting from rest, in which the speed undergoes the same change in every equal interval of time (called uniformly accelerated motion), corresponds to traversing distances that are proportional to the squares of the elapsed times. Then Galileo showed by an experiment that this law is exemplified by motion on an inclined plane. From these two results, Galileo reasoned that the motion of free fall is a case of such uniformly accelerated motion. In the absence of any air resistance, the motion of a freely falling body will always be accelerated according to this law. When Robert Boyle, some thirty years later, was able to evacuate a cylinder, he showed that in such a vacuum all bodies fall with identical speeds no matter what their shapes. Thus proof was given of Galileo's assertion—an extrapolation from experience—that but for air resistance, all bodies fall at the same rate, with the same acceleration. Hence, the speed of a falling body, except for the usually negligible factor of air resistance, depends only on the length of time during which it falls, and not on its weight or the force moving it, as Aristotle had supposed. Today we know the correct value of the acceleration of free falling (sometimes known as the "acceleration of gravity") to be about 32 feet per second change of speed in each second.

Galileo's supreme achievement was not merely to prove that Aristotle had erred and to discover that all bodies, save for the factor of air resistance, fall together despite their differences in weight; others before Galileo had observed this phenomenon. No, what was original with Galileo and revolutionary in its implications was the discovery of the laws of falling bodies and the introduction of a method that combined logical deduction, mathematical analysis, and experiment.

GALILEO'S PREDECESSORS

If we are to appreciate the stature of Galileo properly, we must measure him alongside his contemporaries and predecessors. When, in the final chapter, we see how Newton depended on Galileo's achievement, we shall gain some comprehension of his historical importance. But at this point we shall see exactly how significant he was by making a more realistic appraisal of his originality than is to be found in most textbooks and in all too many histories.

Recall that it was a feature of late Greek (Alexandrian and Byzantine) physics to criticize Aristotle rather than to accept his every word as if it were absolute truth. The same critical spirit characterized Islamic scientific thought and the writings of the medieval Latin West. Thus Dante, whose works are often held to be the acme of medieval European culture, criticized Aristotle for believing "that there were no more than eight heavens [spheres]" and that "the heaven [sphere] of the sun came next after that of the moon, that is, that it was the second from us."

Scholars subjected the Aristotelian law of motion to various corrections, of which the chief features were: (1) concentration on the gradual stages by which motion changes, i.e., acceleration; (2) recognition that in describing changing motion, one can speak only of the speed at some given instant; (3) careful definition of uniform motion—a condition described in a summary of 1369 (by John of Holland) as one in which "the body traverses an equal space in every equal part of the time" *(in omni parte equali temporis)* (which contradicts Galileo's statement on page 89. that he was the first so to define uniform motion); (4) recognition that accelerated motion could be of either a uniform or a nonuniform kind, as diagramed in the following schema:

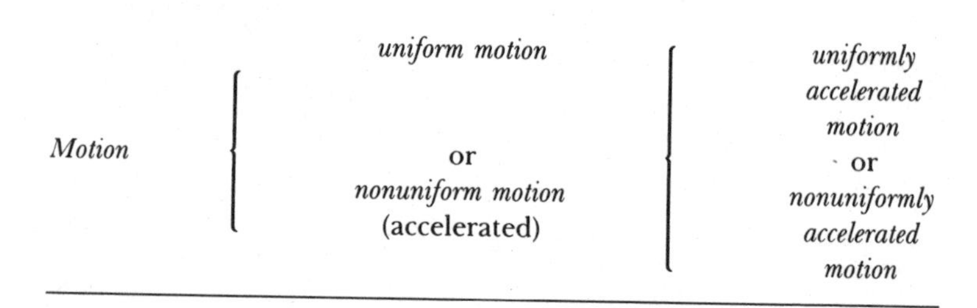

In his presentation Galileo went through this very type of analysis. The simplest motion, he said, is uniform (which he defined in the manner of the scholastics of the fourteenth century); next comes accelerated motion, which may be either uniformly accelerated or nonuniformly accelerated. He chose the simpler, and then explored whether the acceleration is uniform with respect to time or to distance.

In considering how speed may change uniformly, the schoolmen of the fourteenth century proved what is sometimes known as the "mean speed rule." It states that the effect (distance) of a uniformly accelerated motion during any time-interval is exactly the same as if during that interval the moving body had been subject to a uniform motion that was the mean of the accelerated motion. Let us now see this rule expressed in symbols. During time T, suppose a body to be uniformly accelerated from some initial speed V_1 to a final speed V_2. How far (D) will it go? To find the answer determine the average speed $\overline{V}$ during the time-interval; then the distance D is the same as if the body had gone at a constant speed $\overline{V}$ during time T, or $D = \overline{V}T$. Furthermore, since the motion is an example of uniform acceleration, the average speed $\overline{V}$ during the time-interval is the mean of the initial and terminal speeds, so that

$$\overline{V} = \frac{V_1 + V_2}{2}$$

This is very nearly the theorem used by Galileo to prove his own law relating distance to time in accelerated motion. How did the men of the fourteenth century prove it? The first proofs were produced in Merton College, Oxford, by a kind of "word algebra," but in Paris Nicole Oresme proved the theorem geometrically, using a diagram (Fig. 19) very much like the one found in the *Two New Sciences.* *

A major difference between Galileo's presentation and Oresme's is that the latter's was couched in terms of any changing "quality" that might be quantified—including such physical "qualities" as speed, displacement, temperature, whiteness, heaviness, etc., but also such nonphysical "qualities" as love, charity, and grace. But there is no instance in which these men of the fourteenth century tested their results as Galileo did in order to see whether they applied to the real world of experience. For these men the logical exercise of proving the "mean speed rule" was of itself a satisfying experience. For instance, the scientists of the fourteenth century, so far as we know, never even explored the possibility that two objects of unequal weight would fall practically together. Yet, although the fourteenth century

*It follows from the equation for the mean speed ($\overline{V}$) that if the initial speed V_1 is zero, corresponding to motion starting from rest, then, for any speed V at time T, $\overline{V} = ½(O + V) = ½V$. Substituting this result in the equation $D = \overline{V}T$ yields $D = ½(V)T$. Since uniformly accelerated motion is by definition a motion in which speed is proportional to time, or $V \propto T$, the relation $D = ½(V)T$ yields $D \propto T^2$, Galileo's result that in uniformly accelerated motion starting from rest, the distance is proportional to the square of the elapsed time. If the constant of proportionality is A (called "the acceleration"), so that $V = AT$, then the equation $D = ½(V)T$ becomes $D = ½(AT)T$ or $D = ½AT^2$. See, further, note on p. 86.

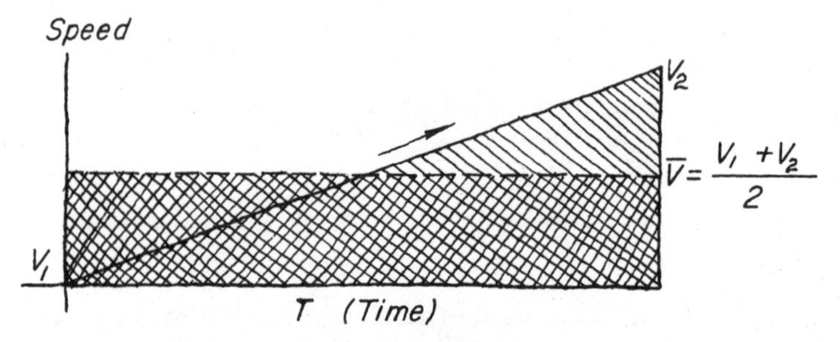

FIG. 19. Nicole Oresme of Paris used geometry to prove that a body uniformly accelerated from an initial speed V_1 to a final speed V_2 would travel the same distance D in the time interval T that it would if it had moved at the constant speed $\overline{V}$, the mean between V_1 and V_2. He assumed that the area under the graph of speed plotted against time would be the distance D. For the uniformly accelerated motion, the graph would be an inclined line and for uniform motion the horizontal line. The area under the first would be the area of a triangle or $1/2T \times V_2$. The area of the second would be the area of the rectangle or $T \times 1/2V_2$, the height of the triangle being twice that of the rectangle. The areas, and therefore the distances traveled, would be equal.

scholastics who discovered the "mean speed rule" did not themselves apply the concept of an acceleration uniform in time to falling bodies as such, one of their successors in the sixteenth century did. By the time of Galileo, the statement that the speed of falling bodies increases continuously as a function of the time had been printed in the book of the Spaniard Domingo de Soto, in which the "mean speed rule" was readily available. But this statement by de Soto appeared as an "aside" and was not presented as a major theorem of nature. It was, to all intents and purposes, buried under a mass of theology and Aristotelian philosophy. (See Supplement 7.)

Another medieval concept of importance in understanding the scientific thought of Galileo is "impetus." This is a property which was supposed to keep things like projectiles moving after they have left the "projector." Impetus resembles both momentum and kinetic energy, and really has no exact equivalent in modern dynamics. It was a distant ancestor of Galileo's concept of inertia and from that developed in turn the modern Newtonian view.*

Galileo's originality was therefore different from what he boastfully declared. No longer need we believe anything so absurd as that there had been no progress in understanding motion between the time of Aristotle and Galileo. And we may ignore the many accounts that make it appear that Galileo invented the modern science of motion in complete ignorance of any medieval or ancient predecessor.

This was a point of view encouraged by Galileo himself, but it is one that could be more justifiably held one hundred years ago than today. One of the most fruitful areas of research in the history of science in the last three quarters of a century—begun chiefly by the French scholar and scientist Pierre Duhem—has been the "exact sciences" of the Middle Ages. These investigations have uncovered a tradition of criticism of Aristotle which paved the way for Galileo's own contributions. By making precise exactly how Galileo advanced beyond his predecessors, we may delineate more accurately his own heroic proportions. In this way, furthermore, we may make the life story of Galileo more real, because we are aware that in the advance of the sciences each man builds on the work of his predecessors. Never was this aspect of the scientific enterprise put better than in the following words of Lord Rutherford (1871–1937), founder of nuclear physics:

> . . . . It is not in the nature of things for any one man to make a sudden violent discovery; science goes step by step, and every man depends on the work of his predecessors. When you hear of a sudden unexpected discovery—a bolt from the blue, as it were—you can always be sure that it has grown up by the influence of one man on another, and it is this mutual influence which makes the enormous possibility of scientific advance. Scientists are not dependent on the ideas of a single man, but on the combined wisdom of thousands of

*Stillman Drake has argued that "medieval natural philosophers adopted impetus theory for their rule of fall, and that excluded the possibility of regarding fall as a case of uniformly difform motion." This is an ingenious explanation of "why nobody ever explicitly raised the question whether speeds varied with time or distance."

> men, all thinking of the same problem, and each doing his little bit to add to the great structure of knowledge which is gradually being erected.

Surely Galileo and Rutherford both typify the spirit of science.

Yet it was Galileo who, for the first time, showed how to resolve the complex motion of a projectile into two separate and different components—one uniform and the other accelerated—and it was Galileo who first put the laws of motion to the test of rigorous experiment and proved that they could be applied to the real world of experience. If it seems that this is only a small achievement, recall that the principles that Galileo made more precise and used as a part of physics rather than a part of logic had been known since the mid-fourteenth century, but that no one else in that 300-year interval had been able to discern how to relate such abstractions to the world of nature. Perhaps in this we may best see the particular quality of his genius in combining the mathematical view of the world with the empirical view obtained by observation, critical experience, and true experiment. (See Supplements 9 and 10.)

FORMULATING THE LAW OF INERTIA

Let us explore a little further Galileo's contribution to scientific methodology in his insistence upon an exact relation between mathematical abstractions and the world of experience. For instance, most of the laws of motion as announced by Galileo would hold true only in a vacuum, where there would be no air resistance. But in the real world it is necessary to deal with the motions of bodies in various kinds of media, in which there is resistance. If Galileo's laws were to be applied to the real world around him, it was necessary for him to know exactly how much effect the resistance of the medium would have. In particular, Galileo was able to show that for bodies with some weight, and not shaped so as to offer enormous resistances to motion through air, the effect of the air was almost negligible. It was this slight factor of air resistance that was responsible for the small difference in the times of descent of light and heavy objects from a given height. This difference was important, because it indicated that air has some resistance, but the smallness of the difference showed how minute the effect of this resistance usually is.

Galileo was able to demonstrate that a projectile follows the path of a parabola because the projectile has simultaneously a combination of two independent motions: a uniform motion in a forward or horizontal direction, and a uniformly accelerated motion downward or in the vertical direction.

Commenting on this result, Galileo has Simplicio quite reasonably say that

> in my opinion it is impossible to remove the impediment of the medium so that this will not destroy the equability of the transverse motion and the rule of acceleration for falling heavy things. All these difficulties make it highly improbable that anything demonstrated from such fickle assumptions can ever be verified in actual experiments.

The reply is then given by Salviati:

> All the difficulties and objections you advance are so well founded that I deem it impossible to remove them. For my part, I grant them all, as I believe our Author would also concede them. I admit that the conclusions demonstrated in the abstract are altered in the concrete, and are so falsified that horizontal [motion] is not uniform; nor does natural acceleration occur [exactly] in the ratio assumed; nor is the line of the projectile parabolic, and so on.

Galileo goes on to prove that

> in projectiles that we find practicable, which are those of heavy material and spherical shape, and even in [others] of less heavy material, and cylindrical shape, as are arrows, launched [respectively] by slings or bows, the deviations from exact parabolic paths will be quite insensible. Indeed I shall boldly say that the smallness of devices usable by us renders external and accidental impediments scarcely noticeable.

In one of his experiments with freely falling bodies Galileo used two balls, one weighing ten or twelve times as much as the other, "one, say, of lead, the other of oak, both descending from a height of 150 or 200 *braccia.*" According to Galileo,

> [E]xperience shows us that two balls of equal size, one of which weighs ten or twelve times as much as the other (for example, one of lead and the other of oak), both descending from a height of 150 or 200 *braccia,* arrive at the earth with very little difference in speed. This assures us that the [role of] the air in impeding and retarding both is small; for if the lead ball, leaving from a height at the same moment as the wooden ball, were but little retarded, and the other a great deal, then over any great distance the lead ball should arrive at the ground leaving the wooden ball far behind, being ten times as heavy. But this does not happen at all; indeed, its victory will not be by even one percent of the entire height; and between a lead ball and a stone ball that weighs one-third or one-half as much, the difference in time of arrival at the ground will hardly be observable.

Next Galileo shows that, apart from weight,

> the impediment received from the air by the same moveable when moved with great speed is not very much more than that which the air opposes to it in slow motion.

He assumed that the resistance which the air offers to the motions under study disturbs "them all in an infinitude of ways, according to the infinitely many ways that the shapes of the moveables vary, and their heaviness, and their speeds." Then he explains:

> As to speed, the greater this is, the greater will be the opposition made to it by the air, which will also impede bodies the more, the less heavy they are. Thus the falling heavy thing ought to go on accelerating in the squared ratio of the duration of its motion; yet, however heavy the moveable may be, when it falls through very great heights the impediment of the air will take away the power of increasing its speed further, and will reduce it to uniform and equable motion. And this equilibration will occur more quickly and at lesser heights as the moveable shall be less heavy.

In this most interesting conclusion, Galileo says that if a body falls long enough, the resistance of the air will increase in some proportion to the speed until the resistance of the air equals and offsets the weight pulling the body down to the earth. If two bodies have the same size, and the same resistance because they have a similar shape, the heavier one will accelerate a longer time, because it has a greater weight. It will continue to accelerate until the resistance (proportional to the speed, which in turn is proportional to the time) equals the weight. What interests us is not this important result so much as Galileo's general conclusion: when the resistance becomes so great that it equals the weight of the falling body, the resistance of the air will prevent any increase in speed and will render the motion uniform. This is to say that if the sum of all the forces acting upon a body (in this case the downward force of the weight and the upward force of the resistance) balances out or equals a net value of zero, the body will nevertheless continue to move, and will move uniformly. This is anti-Aristotelian, because Aristotle held that when the motive force equals the resistance the speed is zero. It is, in limited form, a statement of Newton's first law of motion, or the principle of inertia. According to this principle, the absence of a net external force permits a body either to move in a straight line at constant speed or to stay at rest, and it thus sets up an equivalence between uniform rectilinear motion and rest, a principle that may be considered one of the major foundations of modern Newtonian physics. (See Supplement 8.)

But is Galileo's principle really the same as Newton's? Observe that in Galileo's statement there is not any reference to a general law of inertia, but only to the particular case of downward motion. This is a limited motion, because it can continue only until the falling object strikes the ground. There is no possibility, for example, of such a motion's continuing uniformly in a straight line without limit, as may be inferred from Newton's more general statement.

In the *Two New Sciences* Galileo approached the problem of inertia chiefly in relation to his study of the path of a projectile, which he wanted to show is a parabola (Fig. 20). Galileo consid-

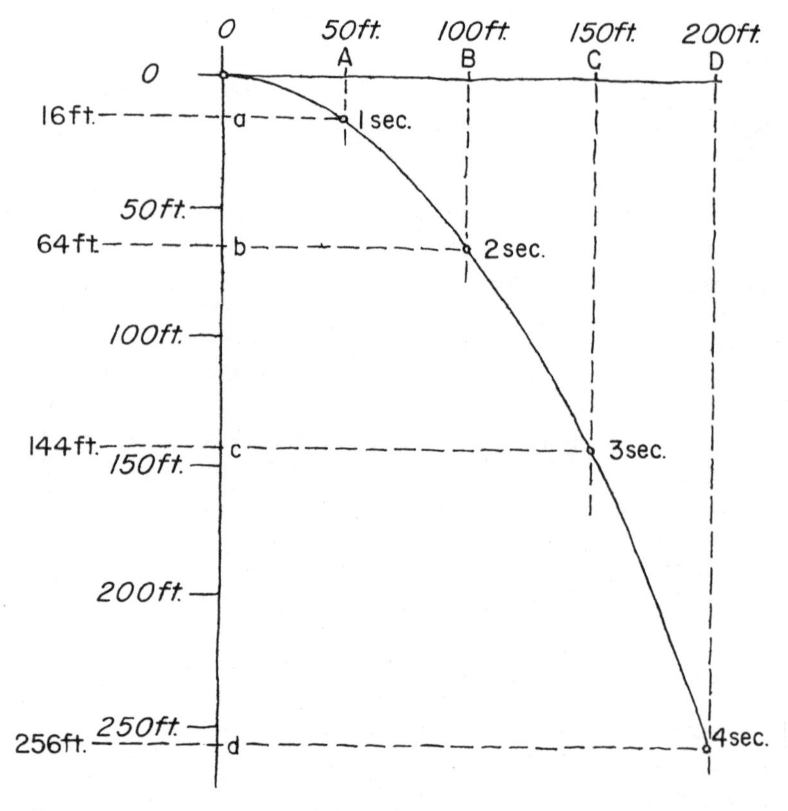

FIG. 20. To see how Galileo analyzed projectile motion, consider a shell fired horizontally from a cannon at the edge of a cliff at a speed of 50 feet per second. The points A, B, C, D show where the shell would be at the ends of successive seconds if there were no air resistance and no downward component, in this case there being a uniform horizontal motion, the shell going 50 feet in each second. In the downward direction, there is an accelerated motion. The points a, b, c, d show where the shell would be if it were to fall with no air resistance and no forward motion. Since the distance is computed by the law

$$D = 1/2AT^2$$

and the acceleration A is 32 ft/sec², the distances corresponding to these times are

T	T^2	$1/2AT^2$	D
1 sec	*1 sec²*	*16 ft/sec² × 1 sec²*	*16 ft.*
2 sec	*4 sec²*	*16 ft/sec² × 4 sec²*	*64 ft.*

ers a body sent out in a horizontal direction. It will then have two separate and independent motions. In the horizontal direction it will move with uniform velocity, except for the small slowing effect of air resistance. At the same time, its downward motion will be accelerated, just as a freely falling body is accelerated. It is the combination of these two motions that causes the trajectory to be parabolic. For his postulate that the downward component of the motion is the same as that of a freely falling body, Galileo did not give an experimental proof, although he indicated the possibility of having one. He devised a little machine in which on an inclined plane (Fig. 21) a ball was projected horizontally, to move in a parabolic path. (See Supplement 9).

Today we can easily demonstrate this conclusion by shooting one of a pair of balls horizontally, while the other is simultaneously allowed to fall freely from the same height. The result of such an experiment is shown in Plate 7, where a series of photographs taken stroboscopically at successive instants shows that although one of the balls is moving forward while the other is dropping vertically, the distances fallen in successive seconds are the same for both. This is the situation of a ball falling on a train

3 sec	*9 sec²*	*16 ft/sec² × 9 sec²*	*144 ft.*
4 sec	*16 sec²*	*16 ft/sec² × 16 sec²*	*256 ft.*

Since the shell actually has the two motions simultaneously, the net path is as shown by the curve.

For those who like a bit of algebra, let v be the constant horizontal speed and x the horizontal distance, so that $x = vt$. In the vertical direction let the distance be y, so that $y = 1/2AT^2$. Then, $x^2 = v^2t^2$ or

$$\left|\begin{array}{l} \dfrac{x^2}{v^2} = t^2 \\[2em] \dfrac{2y}{A} = t^2 \end{array}\right.$$

and $\dfrac{x^2}{v^2} = \dfrac{2y}{A}$ or $y = \dfrac{A}{2v^2}x^2$ which is of the form $y = kx^2$ where k is a constant, and this is the classic equation of the parabola.

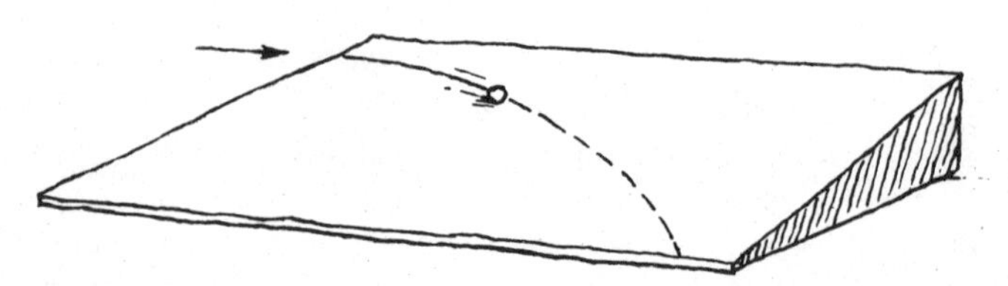

FIG. 21. Galileo's simple apparatus for demonstrating projectile motion was a wedge. A ball started with horizontal motion at the top of the wedge falls toward the bottom of the inclined plane in a parabolic path.

moving at constant speed along a linear track. It falls *vertically* second after second just as it would if the train were at rest. Since it also moves horizontally at the same uniform speed as the train, its true path with respect to the earth is a parabola. Yet another modern example is that of an airplane flying horizontally at constant speed and releasing a bomb or torpedo. The downward fall is the same as if the bomb or torpedo had been dropped from the same height from an object at rest, say a captive balloon on a calm day. As the bomb or torpedo falls from the airplane, it will continue to move forward with the horizontal uniform speed of the airplane and will, except for the effects of the air, remain directly under the plane. But to an observer at rest on the earth, the trajectory will be a parabola.

Finally consider a stone dropped from a tower. With respect to the earth (and for such a short fall the movement of the earth can be considered linear and uniform), it falls straight downward. But with respect to the space determined by the fixed stars, it retains the motion shared with the earth at the moment of release, and its trajectory is therefore a parabola.

These analyses of parabolic trajectories are all based on the Galilean principle of separating a complex motion into two motions (or components) at right angles to each other. It is certainly a measure of his genius that he saw that a body could simultaneously have a uniform or nonaccelerated horizontal component of velocity and an accelerated vertical component—neither one in any way affecting the other. In every such case, the horizontal

PLATE V. "In fair and grateful exchange," as Galileo put it, the earth contributes illumination to the moon. This photograph, taken at Yerkes Observatory, shows earthshine on the portion of the moon that otherwise would be in shadow.

PLATE VI. A ball fired with a spring gun in the smokestack of a moving toy locomotive describes a parabola and lands on the locomotive instead of going straight up and down as it does when the locomotive is standing still. These stroboscopic pictures, with exposures at intervals of one-thirtieth of a second, vividly illustrate one of Galileo's arguments on the behavior of falling bodies and settle the ancient debate about bodies dropped from the masts of moving ships. If the speed of the locomotive were absolutely uniform and if the ball met no air resistance, it would land in the smokestack. (In fact, even under the imperfect conditions of the experiment, the ball hits the smokestack more often than not.) Note that the ball attains the same height whether the locomotive is at rest or moving. Notice, too, that in the picture where the locomotive is standing still, the distances traveled by the ball in the intervals of exposure correspond almost exactly. On the ascent, gravity slows it down; on the descent, gravity speeds it up. Photographs by Berenice Abbott.

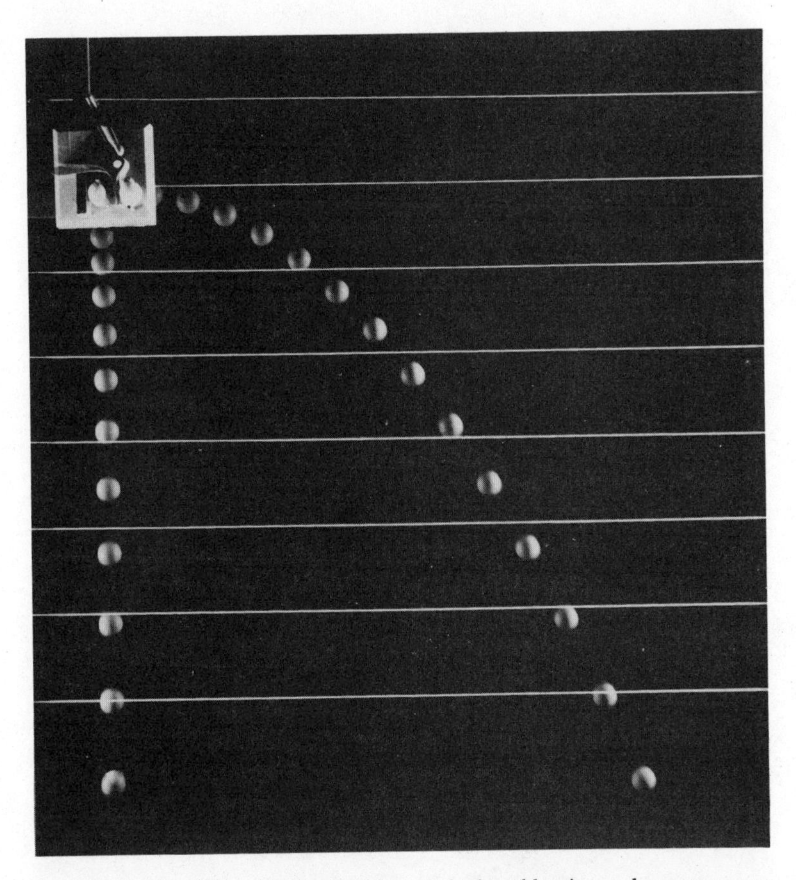

PLATE VII. The independence of the vertical and horizontal components of projectile motion is illustrated in this stroboscopic photograph. In intervals of one-thirtieth of a second the projected ball falling along a parabolic path drops exactly the same distance as the ball allowed to fall vertically. Photograph by Berenice Abbott.

PLATE VIII.

component exemplifies the tendency of a body that is moving at constant speed in a straight line to continue to do so, even though it loses physical contact with the original source of that uniform motion. This may also be described as a tendency of any body to resist any change in its state of motion, a property generally known since Newton's day as a body's inertia. Because inertia is so obviously important for understanding motion, we shall inquire a little more deeply into Galileo's views—not so much to show his limitations as to illustrate how difficult it was to formulate the full law of inertia and to overthrow the last vestiges of the old physics.

But first we may note that in the analysis of the parabolic trajectory, Galileo has departed from a strict kinematics and has introduced some considerations of dynamics. The reason why there is an acceleration in the vertical component of the motion but not in the horizontal component is that gravity acts vertically and not horizontally. Galileo did not conceive of forces as abstractions, and he did not generalize the principles he used in analyzing the motions of projectiles so as to discover a qualitative version of Newton's second law. But later scientists saw in this part of his work the seeds of dynamics. (For a summary of Galileo's achievements in the science of motion, see Supplement 10.)

GALILEAN DIFFICULTIES AND ACHIEVEMENTS: THE LAW OF INERTIA

Toward the end of Galileo's *Two New Sciences* he introduces the subject of projectile motion as follows:

> I mentally conceive of some moveable projected on a horizontal plane, all impediments being put aside. Now it is evident from what has been said elsewhere at greater length that equable [i.e., uniform] motion on this plane would be perpetual if the plane were of infinite extent.

But in Galileo's world of physics, can there be a plane "of infinite extent"? In the real world, one certainly never finds such a plane.

In discussing motion along a plane, Galileo admits the difficulties raised by Simplicio:

> One [of these difficulties] is that we assume the [initial] plane to be horizontal, which would be neither rising nor falling, and to be a straight line—as if every part of such a line could be at the same distance from the center, which is not true. For as we move away from its midpoint towards its extremities, this [line] departs ever farther from the center [of the earth], and hence it is always rising.

Thus, if a ball is moving along any considerable plane tangent to the surface of the earth, this ball will begin to go uphill, which would destroy the uniformity of its motion. But in the real world of experiments, things are different, for then Galileo states that

> in using our instruments, the distances we employ are so small in comparison with the great distance to the center of our terrestrial globe that we could treat one minute of a degree at the equator as if it were a straight line, and two verticals hanging from its extremities as if they were parallel.

Galileo explains what considering an arc a straight line will mean:

> Here I add that we may say that Archimedes and others imagined themselves, in their theorizing, to be situated at infinite distance from the center. In that case their said assumptions would not be false, and hence their conclusions were drawn with absolute proof. Then if we wish later to put to use, for a finite distance [from the center], these conclusions proved by supposing immense remoteness [therefrom], we must remove from the demonstrated truth whatever is significant in [the fact that] our distance from the center is not really infinite, though it is such that it can be called immense in comparison with the smallness of the devices employed by us.

As in his discussion of air resistance, Galileo here wants to know just what the effect may be of a factor that he wishes to ignore. How much error arises from considering a small portion of the earth to be a plane? Very little for most problems.

Earlier, in presenting Galileo's thought on terminal velocities, we called attention to his view that the air resistance increases as some function of the speed. Hence, after falling for some time, a body may generate an air resistance equal to its weight, and then undergo no further acceleration. Under a zero net external force, the body will move in a straight line at constant speed. This is a clear illustration of how a vertical downward motion toward the earth may exemplify a principle of inertia. The projectile seemed, likewise, to exemplify the principle of inertia in its horizontal movement, the component of velocity along the earth. But now we are told that if horizontal motion means motion along a plane tangent to the earth, this motion cannot truly be inertial, since in any direction away from the point of tangency the body, though still moving along the plane, will be going uphill! Evidently, we must accept the conclusion that if such a motion is to be inertial and continue at constant speed *without an external force,* the "plane" on which the body is moving is not a true geometric plane at all but a portion of the earth's surface, which can be taken as planar only because of the relatively large radius of the earth. For Galileo, it would seem the principle of inertia was limited; it was restricted to objects either moving downward along straight line segments terminating at the earth's surface or along small areas on the earth's surface itself. Because the latter motion is not truly along a straight line, Galileo's concept is sometimes referred to as a kind of "circular inertia." But this is unjustified, since it attributes to Galileo a false principle: there is no kind of "inertia" that by and of itself, and without the mediation of something else, can keep a body in constant circular motion.

For enlightenment on Galileo's point of view, we may turn to his *Dialogue Concerning the Two Chief World Systems.* In this work he writes unambiguously of motion that we would call inertial in terms of a circular rather than a linear principle. Here—as in the *Two New Sciences*—he discusses a motion compounded of two separate and independent movements: uniform motion in a circle and accelerated motion in a straight line toward the center of the earth. The reason that Galileo thought in terms of a nonlinear kind of inertia appears to be a desire to explain how on a rotating earth a falling body will always continue to fall downward just as if the earth were at rest. Evidently the straight downward falling of a weight on a rotating earth implied to Galileo that the falling weight must continue to rotate with the earth. Thus he conceived that a ball falling from a tower would continue to move through equal circular arcs in equal times (as any point on the earth does) while nevertheless descending according to the law of uniformly accelerated bodies toward the center of the earth.

There is one place in the *Dialogue* when it almost appears that Galileo has expressed the principle of inertia. Salviati asks Simplicio what would happen to a ball placed on a downward sloping plane. Simplicio agrees that it would accelerate spontaneously. Similarly, on an upward slope, a force would be needed to "thrust it along or even to hold it still." What would happen if such a body were "placed upon a surface with no slope upward or downward?" Simplicio says there would be neither a "natural tendency toward motion" nor a "resistance to being moved." Hence, the object would remain stationary, or at rest. Salviati agrees that this is what would happen if the ball were laid down gently, but if it were given a push directing it toward any part, what would happen? Simplicio replies that it would move in that direction, and that there would not be "cause for acceleration or deceleration, there being no slope upward or downward." There is no cause for "the ball's retardation," nor "for its coming to rest." Salviati then asks how far the ball would continue to move in these circumstances. The reply is, "as far as the extension of the surface continued without rising or falling." Next Salviati says, "Then if such a space were unbounded, the motion on it would likewise be boundless? That is, perpetual?" To which Simplicio assents.

At this point it might seem that Galileo has postulated the modern form of the principle of inertia, in which a body projected on an infinite plane would continue to move uniformly forever. And this is emphasized when Simplicio says that the motion would be "perpetual" if "the body were of durable material." But Salviati then asks him what he thinks is "the cause of the ball moving spontaneously on the downward inclined plane, but only by force on the one tilted upward?" Simplicio replies that "the tendency of heavy bodies is to move toward the center of the earth, and to move upward from its circumference only with force," being put into violent motion. Salviati next says, "Then in order for a surface to be [sloping] neither downward nor upward, all its parts must be equally distant from the center. Are there any such surfaces in the world?" Simplicio replies, "Plenty of them; such would be the surface of our terrestrial globe if it were smooth, and not rough and mountainous as it is. But there is that of water, when it is placid and tranquil." Salviati next says that, accordingly, "a ship, when it moves over a calm sea, is one of these movables which courses over a surface that is tilted neither up nor down, and if all external and accidental obstacles were removed, it would thus be disposed to move incessantly and uniformly from an impulse once received?" Simplicio agrees, "It seems that it ought to be [so]."

Clearly, then, what has seemed at first to be an infinite plane has shrunk in the discussion to a segment of the spherical surface of the earth. And that motion which was said to be "perpetual," and appeared to be uniform motion along an infinite plane, has turned out to be a ship moving on a calm sea, or any other object that moves along a smooth sphere like the earth. And it is precisely this point which Galileo wished to prove, because he now can explain that a stone let fall from a ship will continue to move around the earth as the ship moves, and so will fall from the top of the mast to the foot of the mast. "Now as to that stone which is on top of the mast. Does it not move, carried by the ship, both of them going along the circumference of the circle about its center? And consequently is there not in it an ineradicable motion, all external impediments being removed? And is not this motion as fast as that of the ship?" Simplicio is allowed to draw his own conclusion: "You mean that the stone, moving with an indelibly impressed motion, is not going to leave the [moving] ship, but will follow it, and finally will fall at the same place where it fell when the ship remained motionless."

One of the reasons why Galileo would have found the principle of inertia in its Newtonian form objectionable is that it implies an infinite universe. The Newtonian principle of inertia says that a body moving without the action of any forces will continue to move forever in a straight line at constant speed, and if it moves forever at a constant speed, it must have the potentiality of moving through a space that is unbounded and unlimited. But Galileo states in his *Dialogue Concerning the Two Chief World Systems* that

"Every body constituted in a state of rest but naturally capable of motion will move when set at liberty only if it has a natural tendency toward some particular place." Hence, a body cannot simply move *away from* a place, but only *toward* a place. He also states unequivocally, "Besides, straight motion being by nature infinite (because a straight line is infinite and indeterminate), it is impossible that anything should have by nature the principle of moving in a straight line; or, in other words, toward a place where it is impossible to arrive, there being no finite end. For nature, as Aristotle well says himself, never undertakes to do that which cannot be done, nor endeavors to move whither it is impossible to arrive." It is thus apparent that when Galileo talks about rectilinear motion, he means motion along a limited portion of a straight line or, as we would put it technically, along a straight line segment. For Galileo, as for his medieval predecessors, motion still means "local motion," a translation from one place to another, a motion to a fixed destination and not a motion that merely continues in some specified direction forever—save for circular motions.

Galileo's first published reference to a kind of inertia appears in his famous *History and Demonstrations Concerning Sunspots and Their Phenomena,* published in Rome in 1613, four years after he began his observations with the telescope. In talking about the rotation of the spots around the sun, he set forth a principle of restricted inertia, holding that an object set on a circular path will continue in that path at constant speed along a circle forever, unless there is the action of an external force. This is what he says:

> For I seem to have observed that physical bodies have physical inclination to some motion (as heavy bodies downward), which motion is exercised by them through an intrinsic property and without need of a particular external mover, whenever they are not impeded by some obstacle. And to some other motion they have a repugnance (as the same heavy bodies to motion upward), and therefore they never move in that manner unless thrown violently by an external mover.
>
> Finally, to some movements they are indifferent, as are these same heavy bodies to horizontal motion, to which they have neither inclina-

> tion (since it is not toward the center of the earth) nor repugnance (since it does not carry them away from that center). And therefore, all external impediments removed, a heavy body on a spherical surface concentric with the earth will be indifferent to rest and to movements toward any part of the horizon. And it will maintain itself in that state in which it has once been placed; that is, if placed in a state of rest, it will conserve that; and if placed in movement toward the west (for example), it will maintain itself in that movement. Thus a ship, for instance, having once received some impetus through the tranquil sea, would move continually around our globe without ever stopping; and placed at rest it would perpetually remain at rest, if in the first case all extrinsic impediments could be removed, and in the second case no external cause of motion were added.

Here we may observe that the continuing motion discussed by Galileo is not circular in general but only circular to the extent of being a circle on the surface of the earth, or on a large spherical surface, concentric with the earth. We have seen that Galileo did not consider a small arc of a terrestrial circle to be notably different from a straight line. Even more important, however, is Galileo's introduction (in the second paragraph just quoted)* of the concept of a "state"—of motion or of rest—that (see Supplement 8) was to become a major concept in the new inertial physics of Descartes and of Newton. The problem is made more complicated by the fact that Galileo was undoubtedly acting in accordance with the general ideas of his time, in which a special place was given to circular motions. This was true not only in the Aristotelian physics but also in the Copernican approach to the universe. Copernicus, echoing a neo-Platonic idea, had said that the universe is spherical "either because that figure is the most perfect . . . or because it is the most capacious [i.e., of all possible solids, a sphere has the largest volume for a given surface area] and therefore best suited for that which is to contain and preserve all things; or again because all the perfect parts of it, namely, sun, moon and stars, are so formed; or because all things tend to assume this shape, as is seen in the case of drops of water and

*Galileo's views on inertial motion are discussed in Winifred L. Wisan's *The New Science of Motion* (1974), pp. 261–63; here one may also find a valuable presentation of the "proto-inertial" principle of such of Galileo's predecessors as Cardano and Benedetti (pp. 149–50, 205, 236–37).

liquid bodies in general if freely formed." Since the earth is spherical, Copernicus asked, "Why then hesitate to grant earth that power of motion natural to its [spherical] shape, rather than suppose a gliding round of the whole universe, whose limits are unknown and unknowable?" Galileo's stress on circles and circular motion can be viewed as a concomitant of his advocacy of the Copernican system.

If Galileo is seen to be a creature of his time, still caught up in the principles of circularity in physics, we may observe the extent to which the general thought patterns of an age can limit men of the greatest genius. And the consequences, in the case of Galileo, are particularly interesting in the context of the present book. We shall call attention to two of them, which will be discussed in the next chapter. First of all, Galileo's attachment to circles for planetary orbits hindered him in accepting the concept of elliptical planetary orbits, the outstanding discovery of his contemporary Kepler, published in 1609 just as Galileo was pointing the telescope heavenwards. Secondly, since Galileo restricted the principle of inertia as he conceived it to rotating bodies and to heavy bodies moving freely upon smooth spheres with the same center as the earth (with the exception of terrestrial objects moving on limited straight line segments), he never achieved a true celestial mechanics. Apparently he did not try to explain the orbital motion of the planets by means of any kind of circularly acting inertial principle, and, as Stillman Drake, the leading American Galileo expert, has well said, Galileo "did not attempt any explanation of the cause of planetary motions, except to imply that if the nature of gravity were known this too might be discovered." This was an achievement reserved for Newton.

We shall see that Newton established an inertial physics that provides a dynamics of celestial bodies as well as of terrestrial objects and in which there is only *linear inertia* and no circular inertia at all. No small part of Newton's genius, in fact, is exhibited in his analysis of orbital planetary motions, making use of an idea he learned from Hooke, that in curvilinear motion there is an inertial component *in the linear sense* combined with a continual falling away from the straight line to the orbital path. Hence, unlike Galileo, Newton showed that motion along a circle is noninertial; thus it requires a force. In uniform circular motion Newton and his contemporary Christiaan Huygens showed that there is an acceleration that is nonuniform, and so of a sort that lay beyond Galileo's ken.

Some scholars have seen the whole of Galileo's scientific career as exemplifying his battle for the Copernican system. Certainly his war against Aristotle and Ptolemy was intended to destroy both the concept of a geostatic universe and the physics based upon it. The telescope enabled him to shake the foundations of Ptolemaic astronomy, and his investigations in dynamics led him to a new viewpoint from which events on a moving earth would have the same appearance as on a stationary earth. Galileo did not really explain how the earth could move, but he was successful in showing why terrestrial experiments such as the dropping of weights can neither prove nor disprove the motion of the earth.

The unity of Galileo's scientific life, combining observational astronomy and mathematical physics, comes from his dedication to a sun-centered universe—a dedication reinforced by almost every major discovery he made in either physics or astronomy. Having been the instrument by which the glorious aspects of the creation in the heavens first had been fully revealed to a mortal, Galileo must have had a special sense of urgency to convert all his fellow men to the true—that is, the Copernican—system of the universe. His conflict with the Roman Catholic Church arose because deep in his heart Galileo was a true believer. There was for him no path of compromise, no way to have separate secular and theological cosmologies. If the Copernican system was true as he believed, then what else could Galileo do but fight with every weapon in his arsenal of logic, rhetoric, scientific observation, mathematical theory, and cunning insight, to make his Church accept a new system of the universe? Alas for Galileo, the time was wrong for the Church to make this change, or so it seemed then, following the Council of Trent and its insistence on the literal interpretation of Scripture. There was no avoiding conflict, and the consequences still echo around us in a never-ending literature of controversy. In the contrast between Galileo's heroic stand when he tried to reform the cosmological basis of orthodox theology and his humbled, kneeling surrender when he disavowed his Copernicanism, we may sense the tremendous forces attendant on the birth of modern science. And we may catch a glimpse of the spirit of this great man as we think of him, after his trial and condemnation, living under a kind of house-arrest or surveillance as Milton saw him in Arcetri, completing his greatest scientific work, *Discourses and Demonstrations Concerning Two New Sciences.* This book was the base from which the next generation of scientists would begin the great exploration of the dynamical principles of a sun-centered universe.

CHAPTER
6

Kepler's Celestial Music

Since Greek times scientists have insisted that Nature is simple. A familiar maxim of Aristotle is, "Nature does nothing in vain, nothing superfluous." Another expression of this philosophy has come down to us from a fourteenth-century English monk and scholar, William of Occam. Known as his "law of parsimony" or "Occam's razor" (perhaps for its ruthless cutting away of the superfluous), it maintains, "Entities are not to be multiplied without necessity." "It is vain to do with more what can be done with fewer" perhaps sums up this attitude. As Newton put it, in the *Principia,* "Nature does nothing in vain, and more causes are in vain when fewer suffice." The reason is that "Nature is simple and does not indulge in the luxury of superfluous causes."

We have seen Galileo assume a principle of simplicity in his approach to the problem of accelerated motion, and the literature of modern physical science suggests countless other examples. Indeed, present-day physics is in distress, or at least in an uneasy state, because the recently discovered nuclear "fundamental particles" exhibit a stubborn disinclination to recognize simple laws. Only a few decades ago physicists complacently assumed that the proton and the electron were the only "fundamental particles" they needed to explain the atom. But now one "fundamental particle" after another has crept into the ranks until it appears that there may possibly be as many of them as there are chemical elements. Confronted with this bewildering array, the average physicist is tempted to echo Alfonso the Wise and bemoan the fact that he was not consulted first.

Anyone who examines Fig. 14 (pp. 46-47) will see at once that neither the Ptolemaic nor the Copernican system was, in any sense of the word, "simple." Today we know why these systems lacked simplicity: restricting celestial motions to circles introduced many otherwise unnecessary curves and centers of motion. If astronomers had used some other curves, notably the ellipse, a smaller number of them would have done the job better. It was one of Kepler's great contributions to astronomy to have found this truth.

THE ELLIPSE AND THE KEPLERIAN UNIVERSE

The ellipse enables us to center the solar system on the true sun rather than some "mean sun" or the center of the earth's orbit as Copernicus did. Thus the Keplerian system displays a universe of stars fixed in space, a fixed sun, and a *single* ellipse for the orbit of each planet, with an additional one for the moon. In actual fact, most of these ellipses, except for Mercury's orbit, look so much like circles that at first glance the Keplerian system seems to be the simplified Copernican system shown on page 47 of Chapter 3: one circle for each planet as it moves around the sun, and another for the moon.

An ellipse (Fig. 22) is not as "simple" a curve as a circle, as will be seen. To draw an ellipse (Fig. 22A), stick two pins or thumbtacks into a board, and to them tie the ends of a piece of thread. Now draw the curve by moving a pencil within the loop of thread so that the thread always remains taut. From the method of drawing the ellipse, the following defining condition is apparent: every point P on the ellipse has the property that the sum of the distances from it to two other points F_1 and F_2, known as the *foci,* is constant. (The sum is equal to the length of the string.) For any pair of foci, the chosen length of the string determines the size and shape of the ellipse, which may also be varied by using one string-length and placing the pins near to, or far from, each other. Thus an ellipse may have a shape (Fig. 22B) with more or less the proportions of an egg, a cigar, or a needle, or may be almost round and like a circle. But unlike the true egg, cigar, or needle, the ellipse must always be symmetrical (Fig. 23) with respect to the axes, one of which (the major axis) is a line

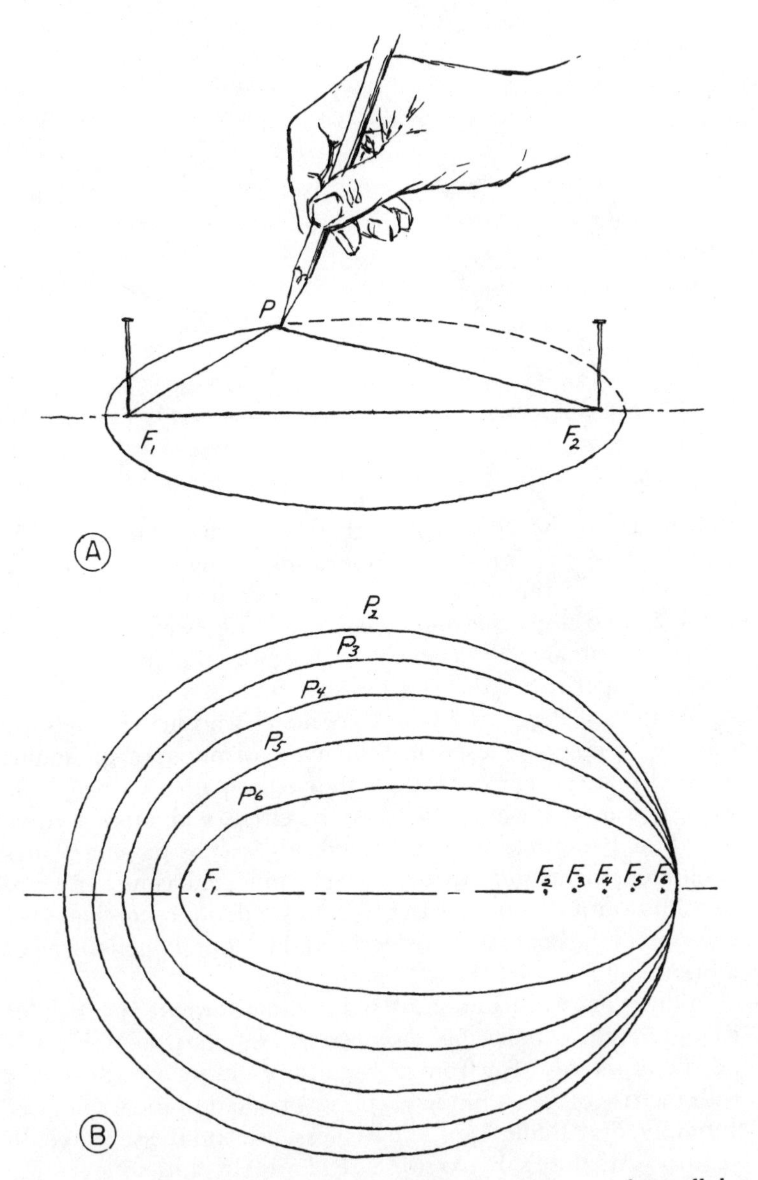

FIG. 22. The ellipse, drawn in the manner shown in (A), can have all the shapes shown in (B) if you use the same string but vary the distance between the pins, as at F_2, F_3, F_4, etc.

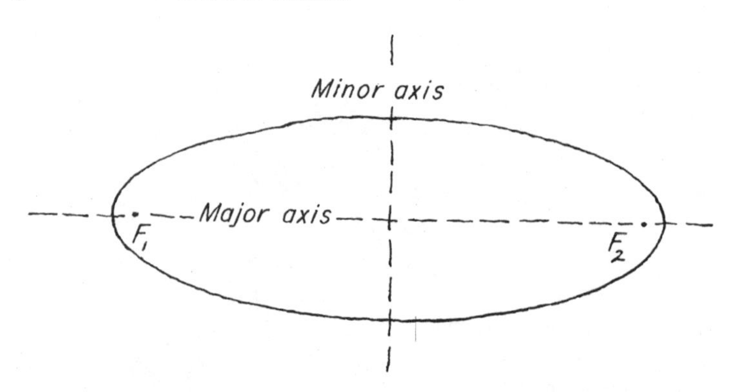

FIG. 23. The ellipse is always symmetrical with respect to its major and minor axes.

drawn across the ellipse through the foci and the other (the minor axis) a line drawn across the ellipse along the perpendicular bisector of the major axis. If the two foci are allowed to coincide, the ellipse becomes a circle; another way of saying this is that the circle is a "degenerate" form of an ellipse.

The properties of the ellipse were described in antiquity by Apollonius of Perga, the Greek geometer who inaugurated the scheme of epicycles used in Ptolemaic astronomy. Apollonius showed that the ellipse, the parabola (the path of a projectile according to Galilean mechanics), the circle, and another curve called the hyperbola may be formed (Fig. 24) by passing planes at different inclinations through a right cone, or a cone of revolution. But until the time of Kepler and Galileo, no one had ever shown that the conic sections occur in the natural phenomena of motion.

In this work we shall not discuss the stages whereby Johannes Kepler came to make his discoveries. Not that the subject is devoid of interest. Far from it! But at present we are concerned with the rise of a new physics, as it was related to the writings of antiquity, the Middle Ages, the Renaissance and the seventeenth century. Aristotle's books were read widely, and so were the writings of Galileo and Newton. Men studied Ptolemy's *Almagest*

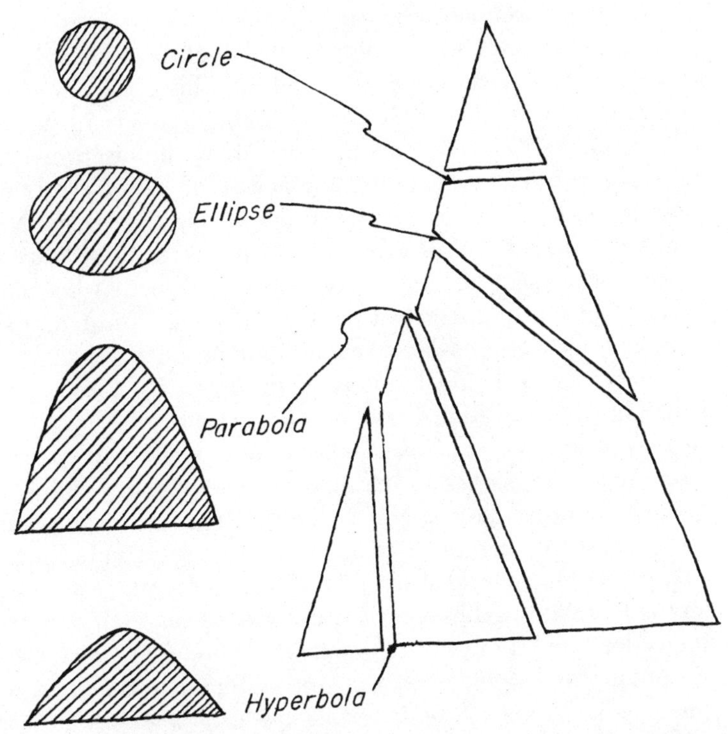

FIG. 24. The conic sections are obtained by cutting a cone in ways shown. Note that the circle is cut parallel to the base of the cone, the parabola parallel to one side.

and Copernicus's *De revolutionibus* carefully. But Kepler's writings were not so generally read. Newton, for example, knew the works of Galileo, but he apparently did not read Kepler's astronomical works. He acquired his knowledge of Kepler's laws at second hand, from T. Streete's handbook of astronomy and V. Wing's textbook. Even today the major works of Kepler are not available in complete English, French, or Italian translations.

This neglect of Kepler's texts is not hard to understand. The language and style are of unimaginable difficulty and prolixity, which, in contrast with the clarity and vigor of Galileo's every word, seem formidable beyond endurance. This is to be expected, for writing reflects the personality of the author. Kepler was a tortured mystic, who stumbled onto his great discoveries in a weird groping that has led one of his biographers* to call him a "sleepwalker." Trying to prove one thing, he discovered another, and in his calculations he made some major errors that canceled each other out. He was utterly unlike Galileo and Newton; never could their purposeful quests for truth conceivably merit the description of sleepwalking. Kepler, who wrote sketches of himself, said that he became a Copernican as a student and that "There were three things in particular, namely, the number, distances and motions of the heavenly bodies, as to which I [Kepler] searched zealously for reasons why they were as they were and not otherwise." About the sun-centered system of Copernicus, Kepler at another time wrote: "I certainly know that I owe it this duty: that since I have attested it as true in my deepest soul, and since I contemplate its beauty with incredible and ravishing delight, I should also publicly defend it to my readers with all the force at my command." But it was not enough to defend the system; he set out to devote his whole life to finding a law or set of laws that would show how the system held together, why the planets had the particular orbits in which they are found, and why they move as they do.

The first installment in this program, published in 1596, when Kepler was twenty-five years old, was entitled *Forerunner of the Dissertations on the Universe, Containing the Mystery of the Universe.* In this book Kepler announced what he considered a great discovery concerning the distances of the planets from the sun. This discovery shows us how rooted Kepler was in the Platonic-Pythagorean tradition, how he sought to find regularities in nature associated with the regularities of mathematics. The Greek geometers had discovered that there are five "regular solids," which are shown in Fig. 25. In the Copernican system there are six planets: Mercury, Venus, Earth, Mars, Jupiter, Saturn. Hence

*Arthur Koestler, *The Sleepwalkers* (London: Hutchinson & Co., 1959).

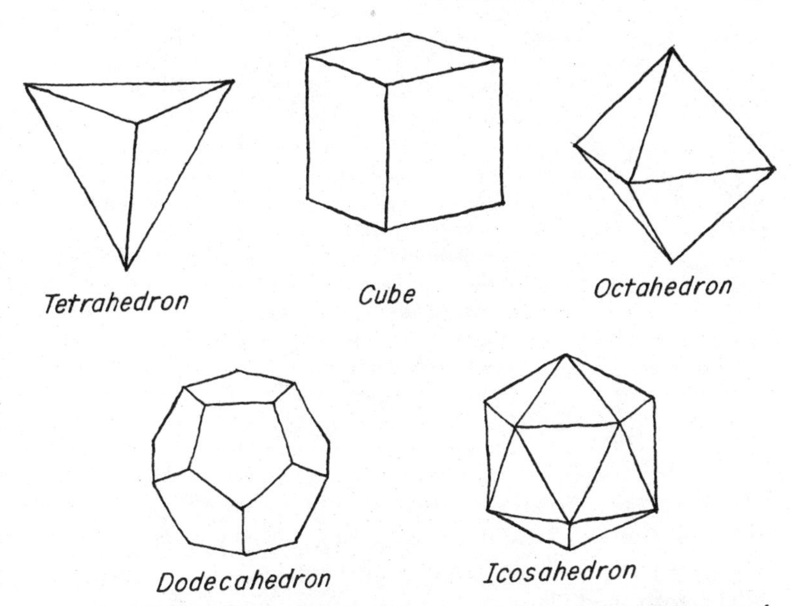

FIG. 25. The "regular" polyhedra. The tetrahedron has four faces, each an equilateral triangle. The cube has six faces, each a square. The octahedron has eight faces, each an equilateral triangle. Each of the dodecahedron's twelve faces is an equilateral pentagon. The twenty faces of the icosahedron are all equilateral triangles.

it occurred to Kepler that five regular solids might separate six planetary orbits.

He started with the simplest of these solids, the cube. A cube can be circumscribed by one and only one sphere, just as one and only one sphere can be inscribed in a cube. Hence we may have a cube that is circumscribed by sphere No. 1 and contains sphere No. 2. This sphere No. 2 just contains the next regular solid, the tetrahedron, which in turn contains sphere No. 3. This sphere No. 3 contains the dodecahedron, which in turn contains sphere No. 4. Now it happens that in this scheme the radii of the successive spheres are in more or less the same proportion as the mean distances of the planets in the Copernican system except for

Jupiter—which isn't surprising, said Kepler, considering how far Jupiter is from the sun. The first Keplerian scheme (Fig. 26), then, was this:

Sphere of Saturn
Cube
Sphere of Jupiter
Tetrahedron
Sphere of Mars
Dodecahedron
Sphere of Earth
Icosahedron
Sphere of Venus
Octahedron
Sphere of Mercury.

"I undertake," he said, "to prove that God, in the creation of this mobile universe and the arrangement of the heavens, had in view the five regular bodies of geometry celebrated since the days of Pythagoras and Plato, and that He has accommodated to their nature, the number of the heavens, their proportions, and the relations of their movements." Even though this book fell short of unqualified success, it established Kepler's reputation as a clever mathematician and as a man who really knew something about astronomy. On the basis of this performance, Tycho Brahe offered him a job.

Tycho Brahe (1546–1601) has been said to have been the reformer of astronomical observation. Using huge and well-constructed instruments, he had so increased the accuracy of naked-eye determinations of planetary positions and of the locations of the stars relative to one another that it was clear that neither the system of Ptolemy nor that of Copernicus could truly predict the celestial appearances. Furthermore, in contrast to earlier astronomers, Tycho did not merely observe the planets now and then to provide factors for a theory or to check such a theory; instead he observed a planet whenever it was visible, night after night. When Kepler eventually became Tycho's successor, he inherited the largest and most accurate collection of planetary observations—notably for the planet

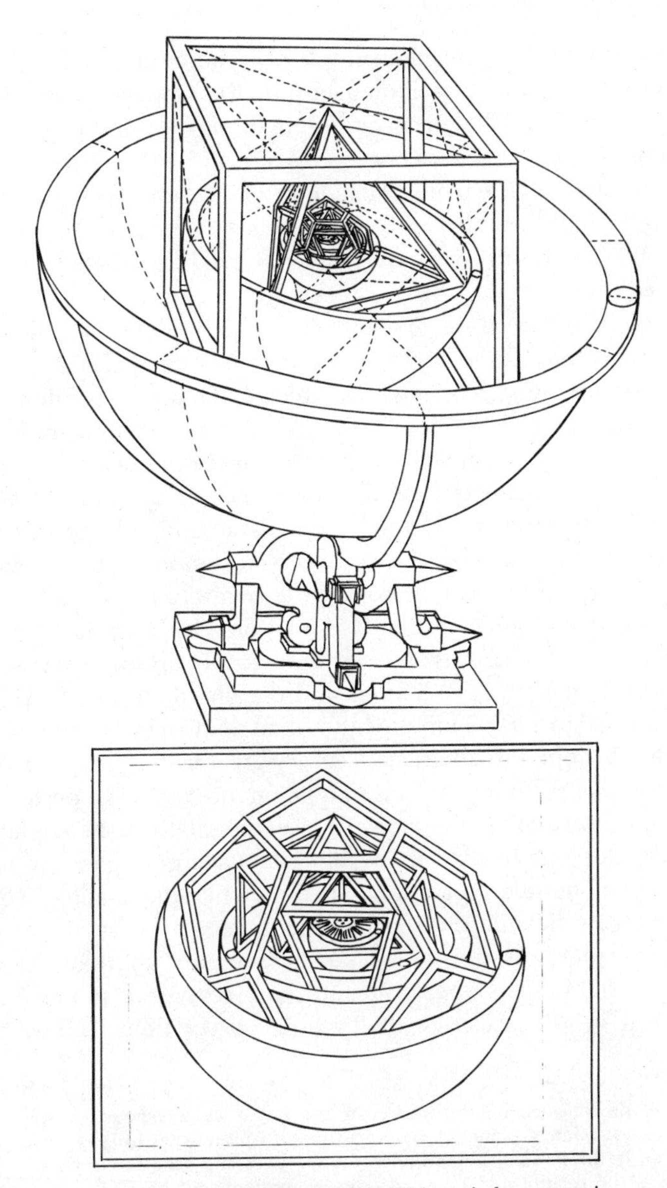

FIG. 26. Kepler's model of the universe. This weird contraption, consisting of the five regular solids fitted together, was dearer to his heart than the three laws on which his fame rests. From his book of 1596.

Mars—that had ever been assembled. Tycho, it may be recalled, believed in neither the Ptolemaic nor the Copernican system but had advanced a geocentric system of his own devising. Kepler, faithful to a promise he had made to Tycho, tried to fit Tycho's data on the planet Mars into the Tychonian system. He failed, as he failed also to fit the data into the Copernican system. But twenty-five years of labor did produce a new and improved theory of the solar system.

Kepler presented his first major results in a work entitled *A New Astronomy . . . Presented in Commentaries on the Motions of Mars,* published in 1609,* the year in which Galileo first pointed his telescope skyward. Kepler had made seventy different trials of putting the data obtained by Tycho into the Copernican epicycles and the Tychonian circles but always failed. Evidently it was necessary to give up all the accepted methods of computing planetary orbits or to reject Tycho's observations as being inaccurate. Kepler's failure may not appear as miserable as he seemed to think. After calculating eccentrics, epicycles, and equants in ingenious combinations, he was able to obtain an agreement between theoretical predictions and the observations of Tycho that was off by only 8 minutes (8′) of angle. Copernicus himself had never hoped to attain an accuracy greater than 10′, and the *Prussian Tables,* computed by Reinhold on the basis of Copernican methods, were off by as much as 5°. In 1609, before the application of telescopes to astronomy, 8′ was not a large angle; 8′ is just twice the minimum separation of two stars that the unaided average eye can distinguish as separate entities.

But Kepler was not to be satisfied by any approximation. He believed in the Copernican sun-centered system and he also believed in the accuracy of Tycho's observations. Thus, he wrote:

*The title indicates that this work is an *Astronomia nova,* a "new astronomy," in the sense of relating planetary motions to their causes so as to be a "celestial physics." In this particular aim Kepler was not successful—the first modern work to reveal the relationship between celestial motions and physical causes was Newton's *Principia* (1687).

> Since the divine goodness has given to us in Tycho Brahe a most careful observer, from whose observations the error of 8′ is shown in this calculation . . . it is right that we should with gratitude recognize and make use of this gift of God. . . . For if I could have treated 8′ of longitude as negligible I should have already corrected sufficiently the hypothesis . . . discovered in chapter xvi. But as they could not be neglected, these 8′ alone have led the way towards the complete reformation of astronomy, and have been made the subject-matter of a great part of this work.

Starting afresh, Kepler finally took the revolutionary step of rejecting circles altogether, trying an egg-shaped oval curve and eventually the ellipse. To appreciate how revolutionary this step actually was, recall that both Aristotle and Plato had insisted that planetary orbits had to be combined out of circles, and that this principle was a feature common to both Ptolemy's *Almagest* and Copernicus's *De revolutionibus.* Galileo, Kepler's friend, politely ignored the strange aberration. But the final victory was Kepler's. He not only got rid of innumerable circles, requiring but one oval curve per planet, but he made the system accurate and found a wholly new and unsuspected relation between the location of a planet and its orbital speed.

THE THREE LAWS

Kepler's problem was not only to determine the orbit of Mars, but at the same time to find the orbit of the earth. The reason is that our observations of Mars are made from the earth, which itself does not move uniformly in a perfect circle around the sun. Fortunately, however, the earth's orbit is almost circular. Kepler discarded Copernicus's idea that all planetary orbits should be centered on the mid-point of the earth's orbit. He discovered, instead, that *the orbit of each planet is in the shape of an ellipse with the sun located at one focus.* This principle is known as Kepler's first law.*

*In his book on Mars, Kepler first derives a general law of areas that is independent of any particular orbit. Only later, and by dint of enormous labor in calculation, does he invent the concept of an elliptical orbit, then finding that the orbit

Kepler's second law tells us about the speed with which a planet moves in its orbit. This law states that *in any equal time intervals, a line from the planet to the sun will sweep out equal areas.* Fig. 27 shows equal areas for three regions in a planetary orbit. Since the three shaded regions are of equal area, the planet moves most quickly when nearest to the sun and most slowly when farthest from the sun. This second law thus tells us at once that the apparent irregularity in the speed with which planets move in their orbits is a variation that is a function of a simple geometric condition.

The first and second laws plainly show how Kepler altered and simplified the Copernican system. But the third law, known also as the harmonic law, is even more interesting. It is called the harmonic law because its discoverer thought it demonstrated the true celestial harmonies. Kepler even entitled the book in which he announced it *The Harmony of the World* (1619). The third law states a relation between the periodic times in which the planets complete their orbits about the sun and their average distances from the sun. Let us make a table of the periodic times (T) and average distances (D). In this table and in the following text, the distances are given in astronomical units. One astronomical unit is, by definition, the mean distance from the earth to the sun. This table shows us that there is no simple relationship between D and T. Kepler therefore tried to see what would happen if he

	Mercury	*Venus*	*Earth*	*Mars*	*Jupiter*	*Saturn*
periodic time T (years)	0.24	0.615	1.00	1.88	11.86	29.457
mean distance from the sun D (astronomical units)	0.387	0.723	1.00	1.524	5.203	9.539

fits the observations re Mars. Some eighty years later, in the *Principia,* Newton deals with the area law first (props. 1–3) and only later (prop. 11) with the law of elliptical orbits.

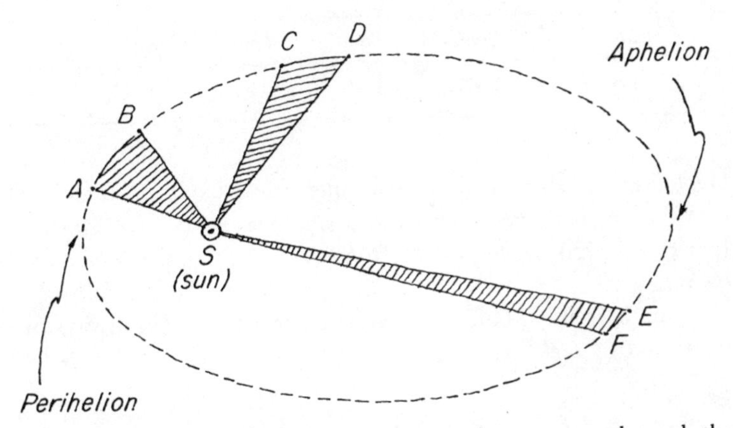

FIG. 27. Kepler's law of equal areas. Since a planet moves through the arcs $\widehat{AB}$, $\widehat{CD}$ and $\widehat{EF}$ in equal times (because the areas SAB, SCD, and SEF are equal), it travels fastest at perihelion, when nearest the sun, and slowest at aphelion, when farthest from the sun. The shape of this ellipse is that of a comet's orbit. Planetary ellipses are more nearly circular.

took the squares of these values, D^2 and T^2. These may be tabulated as follows (using today's values):

	Mercury	*Venus*	*Earth*	*Mars*	*Jupiter*	*Saturn*
T^2	0.058	0.378	1.00	3.53	141	867.7
D^2	0.150	0.523	1.00	2.323	27.071	90.993

There is still no relation discernible between D and T^2, or between D^2 and T, or even between D^2 and T^2. Any ordinary mortal would have given up at this point. Not Kepler! He was so convinced that these numbers must be related that he would never have given up. The next power is the cube. T^3 turns out to be of no use, but D^3 yields the following numbers. Note them and then turn back to the table of squares.

	Mercury	*Venus*	*Earth*	*Mars*	*Jupiter*	*Saturn*
D^3	0.058	0.378	1.00	3.54	141	867.9

Here then are the celestial harmonies, the third law, which states that the *squares of times of revolution of any two planets around the sun* (earth included) *are proportional to the cubes of their mean distances from the sun.*

In mathematical language, we may say that "T^2 is always proportional to D^3" or

$$\frac{D^3}{T^2}=K,$$

where K is a constant. If we choose as units for D and T the astronomical unit and the year, then K has the numerical value of unity. (But if the distance were measured in miles and time in seconds, the value of the constant K would not be unity.) Another way of expressing Kepler's third law is

$$\frac{D_1{}^3}{T_1{}^2}=\frac{D_2{}^3}{T_2{}^2}=\frac{D_3{}^3}{T_3{}^2}=\frac{D_4{}^3}{T_4{}^2}=\ldots\ldots\ldots=K$$

where D_1 and T_1, D_2 and T_2, . . . , are the respective distances and periods of any planet in the solar system.

To see how this law may be applied, let us suppose that a new planet were discovered at a mean distance of $4AU$ from the sun. What is its period of revolution? Kepler's third law tells us that the ratio D^3/T^2 for this new planet must be the same as the ratio $D_0{}^3/T_0{}^2$ for the earth. That is,

$$\frac{D^3}{T^2}=\frac{(1AU)^3}{(1^y)^2}.$$

Since $D = 4AU$,

$$\frac{(4AU)^3}{T^2}=\frac{(1AU)^3}{(1^y)^2},$$

$$\frac{64}{T^2} = \frac{1}{(1^y)^2}$$

$$T^2 = 64 \times (1^y)^2$$

$$T = 8^y.$$

The inverse problem may also be solved. What is the distance from the sun of a planet having a period of 125 years?

$$\frac{D^3}{T^2} = \frac{(1AU)^3}{(1^y)^2}$$

$$\frac{D^3}{(125^y)^2} = \frac{(1AU)^3}{(1^y)^2}$$

$$\frac{D^3}{125 \times 125} = \frac{(1AU)^3}{1}$$

$$D^3 = 25 \times 25 \times 25 \times (1AU)^3$$

$$D = 25AU.$$

Similar problems can be solved for any satellite system. The significance of this third law is that it is a law of necessity; that is, it states that it is impossible in any satellite system for satellites to move at just any speed or at any distance. Once the distance is chosen, the speed is determined. In our solar system this law implies that the sun provides the governing force that keeps the planets moving as they do. In no other way can we account for the fact that the speed is so precisely related to distance from the sun. Kepler thought that the action of the sun was, in part at least, magnetic. It was known in his day that a magnet attracts another magnet even though considerable distances separate them. The motion of one magnet produces motion in another. Kepler was aware that a physician of Queen Elizabeth, William Gilbert (1544–1603), had shown the earth to be a huge magnet. If all objects in the solar system are alike rather than different, as Galileo had shown and as the heliocentric system implies, why should not the sun and the other planets also be magnets like the earth?

Kepler's supposition, however tempting, does not lead directly to an explanation of why planets move in ellipses and sweep out equal areas in equal times. Nor does it tell us why the particular distance-period relation he found actually holds. Nor does it seem in any way related to such problems as the downward fall of bodies—according to the Galilean law of fall—on a stationary or on a moving earth, since the average rock or piece of wood is not magnetic. And yet we shall see that Newton, who eventually answered all these questions, based his discoveries on the laws found by Kepler and Galileo.

KEPLER VERSUS THE COPERNICANS

Why were Kepler's beautiful results not universally accepted by Copernicans? Between the time of their publication (I, II, 1609; III, 1619) and the publication of Newton's *Principia* in 1687, there are very few works that contain references to all three of Kepler's laws. Galileo, who had received copies of Kepler's books and who was certainly aware of the proposal of elliptic orbits, never referred in his scientific writings to any of the laws of Kepler, either to praise or to criticize them. In part, Galileo's reaction must have been Copernican, to stick to the belief in true circularity, implied in the very title of Copernicus's book: *On the Revolutions of the Celestial Spheres.* That work opened with a theorem: 1. *That the Universe is Spherical.* This is followed shortly after by a discussion of the topic, "That the motion of the heavenly bodies is uniform, circular, and perpetual, or composed of circular motions." The main line here is:

> Rotation is natural to a sphere and by that very act is its shape expressed. For here we deal with the simplest kind of body, wherein neither beginning nor end may be discerned nor, if it rotates ever in the same place, may the one be distinguished from the other. . . .
>
> We must conclude [despite any observed apparent irregularities, such as the retrogradations of planets] that the motions of these bodies are ever circular or compounded of circles. For the irregularities themselves are subject to a definite law and recur at stated times, and this could not happen if the motions were not circular, for a circle alone can thus restore the place of a body as it was. So with the Sun

which, by a compounding of circular motions, brings ever again the changing days and nights and the four seasons of the year.

Kepler thus was acting in a most un-Copernican way by not assuming that the planetary orbits are either "circles" or "compounded of circles"; furthermore, he had come to his conclusion in part by reintroducing, at one stage of his thought, the one aspect of Ptolemaic astronomy to which Copernicus had most objected, the *equant.* In his astronomy, Kepler introduced a simple approximation to take the place of the law of areas. Kepler said that a line from any planet to the empty focus of its ellipse (Fig. 28) rotates uniformly, or that it does so very nearly. The empty focus, or the point about which such a line would rotate through equal angles in equal times, is the equant. (Incidentally, we may observe that this latter "discovery" of Kepler's is not true.)

From almost every point of view, the ellipses must have seemed objectionable. What kind of force could steer a planet along an

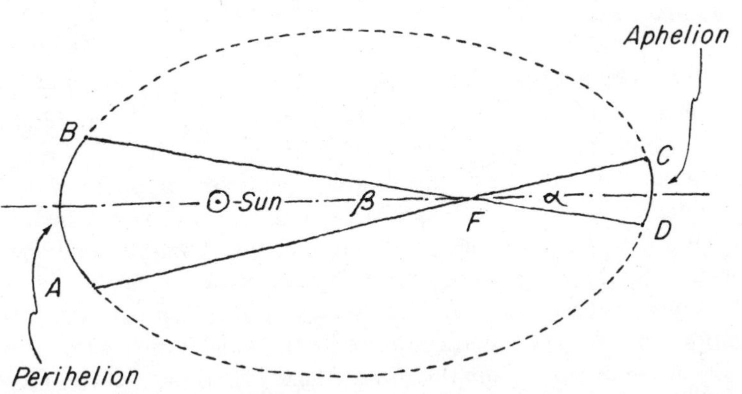

FIG. 28. Kepler's law of the equant. If a planet moves so that in equal times it sweeps out equal angles with respect to the empty focus at F, it will move through arcs $\widehat{AB}$ and $\widehat{CD}$ in the same time because the angles α and β are equal. According to this law, the planet moves faster along arc $\widehat{AB}$ (at perihelion) than along arc $\widehat{CD}$ (at aphelion) as the law of equal areas predicts. Nevertheless, this law is only a rough approximation. But in the seventeenth century, certain correction factors were added to this law to make it give more accurate results.

elliptical path with just the proper variation of speed demanded by the law of equal areas? We shall not reproduce Kepler's discussion of this point, but shall confine our attention to one aspect of it. Kepler supposed that some kind of force or emanation comes out of the sun and moves the planets. This force—it is sometimes called an *anima motrix*—does not spread out in all directions from the sun. Why should it? After all, its function is only to move the planets, and the planets all lie in, or very nearly in, a single plane, the plane of the ecliptic. Hence Kepler supposed that this *anima motrix* spread out only in the plane of the ecliptic. Kepler had discovered that light, which spreads in all directions from a luminous source, diminishes in its intensity as the inverse square of the distance; that is, if there is a certain intensity or brightness three feet away from a lamp, the brightness six feet away will be one-fourth as great because four is the square of two and the new distance is twice the old. In equation form,

$$\text{intensity} \propto \frac{1}{(\text{distance})^2}$$

But Kepler held that the solar force does not spread out in all directions according to the inverse-square law, as the solar light does, but only in the plane of the ecliptic according to a quite different law. It is from this doubly erroneous supposition that Kepler derived his law of equal areas—and he did so *before* he had found that the planetary orbits are ellipses! The difference between Kepler's procedure and what we would consider to be "logical" is that Kepler did *not* first find the actual path of Mars about the sun, and then compute its speed in terms of the area swept out by a line from the sun to Mars. This is but one example of the difficulty in following Kepler through his book on Mars.

THE KEPLERIAN ACHIEVEMENT

Galileo particularly disliked the idea that solar emanations or mysterious forces acting at a distance could affect the earth or any part of the earth. He not only rejected Kepler's suggestion that the sun might be the origin of an attractive force moving the earth and planets (on which the first two laws of Kepler were based), but he especially rejected Kepler's suggestion that a lunar force or emanation might be a cause of the tides. Thus he wrote:

> But among all the great men who have philosophized about this remarkable effect, I am more astonished at Kepler than at any other. Despite his open and acute mind, and though he has at his fingertips the motions attributed to the earth, he has nevertheless lent his ear and his assent to the moon's dominion over the waters, and to occult properties, and to such puerilities.

As to the harmonic law, or third law, we may ask with the voice of Galileo and his contemporaries, Is this science or numerology? Kepler already had committed himself publicly to the belief that the telescope should reveal not only the four satellites of Jupiter discovered by Galileo, but two of Mars and eight of Saturn. The reason for these particular numbers was that then the number of satellites per planet would increase according to a regular geometric sequence: 1 (for the earth), 2 (for Mars), 4 (for Jupiter), 8 (for Saturn). Was not Kepler's distance-period relation something of the same pure number-juggling rather than true science? And was not evidence for the generally nonscientific aspect of Kepler's whole book to be found in the way he tried to fit the numerical aspects of the planets' motions and locations into the questions posed by the table of contents for Book Five of his *Harmony of the World?*

1. Concerning the five regular solid figures.
2. On the kinship between them and the harmonic ratios.
3. Summary of astronomical doctrine necessary for contemplation of the celestial harmonies.
4. In what things pertaining to the planetary movements the simple harmonies have been expressed and that all those harmonies which are present in song are found in the heavens.
5. That the clefs of the musical scale, or pitches of the system, and the kinds of harmonies, the major and the minor, are expressed by certain movements.

6. That each musical Tone or Mode is in a certain way expressed by one of the planets.
7. That the counterpoints or universal harmonies of all the planets can exist and be different from one another.
8. That the four kinds of voice are expressed in the planets: soprano, alto, tenor, and bass.
9. Demonstration that in order to secure this harmonic arrangement, those very planetary eccentricities which any planet has as its own, and no others, had to be set up.
10. Epilogue concerning the sun, by way of very fertile conjectures.

Below are shown the "tunes" played by the planets in the Keplerian scheme.

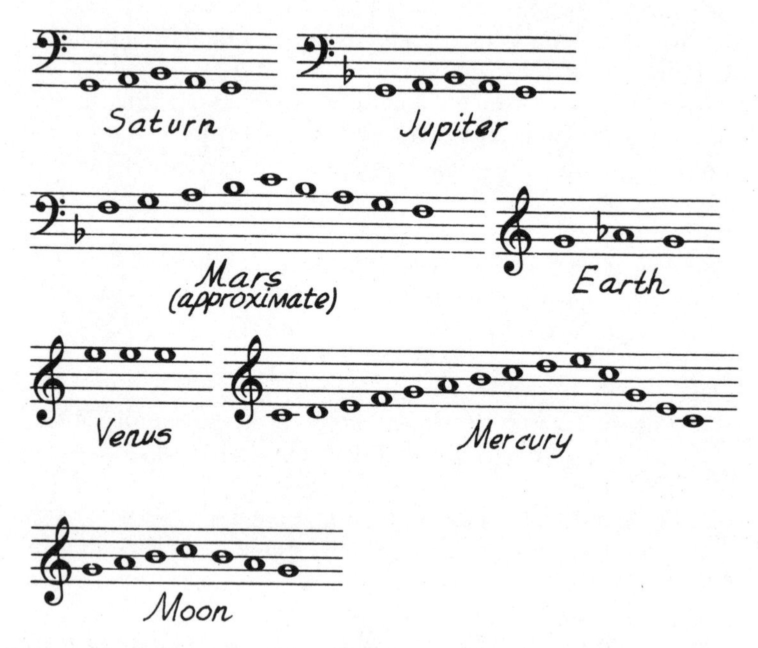

FIG. 29. Kepler's music of the planets, from his book *The Harmony of the World.* Small wonder a man of Galileo's stamp never bothered to read it!

Surely a man like Galileo would find it hard to consider such a book a serious contribution to celestial physics.

Kepler's last major book was an *Epitome of Copernican Astronomy,* completed for publication nine years before his death in 1630. In it he defended his departures from the original Copernican system. But what is of the most interest to us is that in this book, as in the *Harmony of the World* (1619), Kepler again proudly presented his earliest discovery concerning the five regular solids and the six planets. It was, he still maintained, the reason for the number of planets being six.

It must have been almost as much work to disentangle the three laws of Kepler from the rest of his writings as to remake the discoveries. Kepler deserves credit for having been the first scientist to recognize that the Copernican concept of the earth as a planet and Galileo's discoveries demanded that there be one physics—applying equally to the celestial objects and ordinary terrestrial bodies. But, alas, Kepler remained so enmeshed in Aristotelian physics that when he attempted to project a terrestrial physics into the heavens, the basis still came essentially from Aristotle. Thus the major aim of Keplerian physics remained unachieved, and the first workable physics for heaven and earth derived not from Kepler but from Galileo and attained its form under the magistral guidance of Isaac Newton.*

*Kepler did introduce the term "inertia" into the physics of motion, but the sense of Keplerian "inertia" was very different from the later (and present) significance of this term; see Supplement 8.

CHAPTER
7

The Grand Design—A New Physics

The publication of Isaac Newton's *Principia* in 1687 was one of the most notable events in the whole history of physical science. In it one may find the culmination of thousands of years of striving to comprehend the system of the world, the principles of force and of motion, and the physics of bodies moving in different media. It is no small testimony to the vitality of Newton's scientific genius that although the physics of the *Principia* has been altered, improved, and challenged ever since, we still set about solving most problems of celestial mechanics and the physics of gross bodies by proceeding essentially as Newton did some 300 years ago. Newtonian principles of celestial mechanics guide our artificial satellites, our space shuttles, and every spacecraft we launch to explore the vast reaches of our solar system. And if this is not enough to satisfy the canons of greatness, Newton was equally great as a pure mathematician. He invented the differential and integral calculus (produced simultaneously and independently by the German philosopher Gottfried Wilhelm Leibniz), which is the language of physics; he developed the binomial theorem and various properties of infinite series; and he laid the foundations for the calculus of variations. In optics, Newton began the experimental study of the analysis and composition of light, showing that white light is a mixture of light of many colors, each having a characteristic index of refraction. Upon these researches have risen the science of spectroscopy and the methods of color analysis. Newton invented a reflecting telescope and so showed astronomers how to transcend the limitations of telescopes built of lenses. All in all, his was a fantastic scientific achievement—of a kind that has never been equaled and may never be equaled again.

In this book we shall deal exclusively with Newton's system of dynamics and gravitation, the central problems for which the preceding chapters have been a preparation. If you have read them carefully, you have in mind all but one of the major ingredients requisite to an understanding of the Newtonian system of the world. But even if that one were to be given—the analysis of uniform circular motion—the guiding hand of Newton would still be required to put the ingredients together. It took genius to supply the new concept of universal gravitation. Let us see what Newton actually did.

First of all, it must be understood that Galileo himself never attempted to display any scheme of forces that would account for the movement of the planets, or of their satellites. As for Copernicus, the *De revolutionibus* contains no important insight into a celestial mechanics. Kepler had tried to supply a celestial mechanism, but the result was never a very happy one. He held that the *anima motrix* emanating from the sun would cause planets to revolve about the sun in circles. He further supposed that magnetic interactions of the sun and a planet would shift the planet during an otherwise circular revolution into an elliptical orbit. Others who contemplated the problems of planetary motion proposed systems of mechanics containing certain features that were later to appear in Newtonian dynamics. One of these was Robert Hooke, who quite understandably thought that Newton should have given him more credit than a mere passing reference for having anticipated parts of the laws of dynamics and gravitation.

NEWTONIAN ANTICIPATIONS

The climactic chapter in the discovery of the mechanics of the universe starts with a pretty story. By the third quarter of the seventeenth century, a group of men had become so eager to advance the new mathematical experimental sciences that they banded together to perform experiments in concert, to present problems for solution to one another, and to report on their own researches and on those of others as revealed by correspondence, books, and pamphlets. Thus it came about that Robert Hooke, Edmond Halley, and Sir Christopher Wren, England's foremost architect, met to discuss the question, Under what law of force would a planet follow an elliptical orbit? From Kepler's laws—especially the third or harmonic law, but also the second or law of areas—it was clear that the sun somehow or other must control or at least affect the motion of a planet in accordance with the relative proximity of the planet to the sun. Even if the particular mechanisms proposed by Kepler (an *anima motrix* and a magnetic force) had to be rejected, there could be no doubt that some kind of planet-sun interaction keeps the planets in their courses. Furthermore, a more acute intuition than Kepler's would sense that any force emanating from the sun must spread out in all directions from that body, presumably diminishing according to the inverse of the square of its distance from the sun—as the intensity of light diminishes in relation to distance. But to say this much is a very different thing from *proving* it mathematically. For to prove it would require a complete physics with mathematical methods for solving all the attendant and consequent problems. When Newton declined to credit authors who tossed off general statements without being able to prove them mathematically or fit them into a valid framework of dynamics, he was quite justified in saying, as he did of Hooke's claims: "Now is not this very fine? Mathematicians that find out, settle, and do all the business must content themselves with being nothing but dry calculators and drudges; and another, that does nothing but pretend and grasp at all things, must carry away all the invention, as well of those that were to follow him as of those that went before." (See, further, Supplement 11).

In any event, by January 1684 Halley had concluded that the force acting on planets to keep them in their orbits "decreased in the proportion of the squares of the distances reciprocally,"

$$F \propto \frac{1}{D^2},$$

but he was not able to deduce from that hypothesis the observed motions of the celestial bodies. When Wren and Hooke met later in the month, they agreed with Halley's supposition of a solar force. Hooke boasted "that upon that principle all the laws of the celestial motions were to be [i.e., could be] demonstrated, and that he himself had done it." But despite repeated urgings and Wren's offer of a considerable monetary prize, Hooke did not—and presumably could not—produce a solution. Six months later, in August 1684, Halley decided to go to Cambridge to consult Isaac Newton. On his arrival he learned the "good news" that Newton "had brought this demonstration to perfection." Here is DeMoivre's almost contemporaneous account of that visit:

> After they had been some time together, the Dr. [Halley] asked him what he thought the curve would be that would be described by the planets supposing the force of attraction towards the sun to be reciprocal to the square of their distance from it. Sir Isaac replied immediately that it would be an ellipsis. The Doctor, struck with joy and amazement, asked him how he knew it. Why, saith he, I have calculated it. Whereupon Dr. Halley asked him for his calculation without any further delay. Sir Isaac looked among his papers but could not find it, but he promised him to renew it and then to send it him. Sir Isaac, in order to make good his promise, fell to work again, but he could not come to that conclusion which he thought he had before examined with care. However, he attempted a new way which, though longer than the first, brought him again to his former conclusion. Then he examined carefully what might be the reason why the calculation he had undertaken before did not prove right, and he found that, having drawn an ellipsis coarsely with his own hand, he had drawn the two axes of the curve, instead of drawing two diameters somewhat inclined to one another, whereby he might have fixed his imagination to any two conjugate diameters, which was requisite he should do. That being perceived, he made both his calculations agree together.

Spurred on by Halley's visit, Newton resumed work on a subject that had commanded his attention in his twenties when he had laid the foundations of his other great scientific discoveries: the nature of white light and color and the differential and integral calculus. He now put his investigations in order, made great progress, and in the fall term of the year, discussed his research in a series of lectures on dynamics that he gave at Cambridge University, as required by his professorship. Eventually, with Halley's encouragement, a draft of these lectures, *De motu cor-*

porum, grew into one of the greatest and most influential books any man has yet conceived. Many a scientist has echoed the sentiment that Halley expressed in the ode he wrote as a preface to Newton's *Principia* (or, to give Newton's masterpiece its full title, *Philosophiae naturalis principia mathematica, Mathematical Principles of Natural Philosophy,* London, 1687):

> *Then ye who now on heavenly nectar fare,*
> *Come celebrate with me in song the name*
> *Of Newton, to the Muses dear; for he*
> *Unlocked the hidden treasuries of Truth:*
> *So richly through his mind had Phoebus cast*
> *The radiance of his own divinity.*
> *Nearer the gods no mortal may approach.*

THE PRINCIPIA

The *Principia* is divided into three parts or books; we shall concentrate on the first and third. In Book One Newton develops the general principles of the dynamics of moving bodies, and in Book Three he applies the principles to the mechanism of the universe. Book Two deals with a facet of fluid mechanics, the theory of waves, and other aspects of physics.

In Book One, following the preface, a set of definitions, and a discussion of the nature of time and space, Newton presented the "axioms, or laws of motion":

Law I

Every body perseveres in its state of being at rest or of moving uniformly straight forward, except insofar as it is compelled to change its state by forces impressed upon it.

Law II

A change in motion is proportional to the motive force impressed and takes place in the direction of the straight line along which that force is impressed. [See Suppl. Note on p. 184.]

Observe that if a body is in uniform motion in a straight line, a force at right angles to the direction of motion of the body will not affect the forward motion. This follows from the fact that the acceleration is always in the same direction as the force producing it, so that the acceleration in this case is at right angles to the direction of motion. Thus in the toy train experiment of Chapter 5, the chief force acting is the downward force of gravity, producing a vertical acceleration. The ball, whether moving forward or at rest, is thus caused to slow down in its upward motion until it comes to rest, and then be speeded up or accelerated on the way down.

A comparison of the two sets of photographs (p. 83) shows that the upward and downward motions are exactly the same whether the train is at rest or in uniform motion. In the forward direction there is no effect of weight or gravity, since this acts only in a downward direction. The only force in the forward or horizontal direction is the small amount of air friction, which is almost negligible; so one may say that in the horizontal direction there is no force acting. According to Newton's first law of motion, the ball will continue to move in the forward direction with uniform motion in a straight line just as the train does—a fact you can check by inspecting the photograph. The ball remains above the locomotive whether the train is at rest or in uniform motion in a straight line. This law of motion is sometimes called the *principle of inertia,* and the property that material bodies have of continuing in a state of rest or of uniform motion in a straight line is sometimes known as the bodies' *inertia.* *

Newton illustrated Law I by reference to projectiles that continue in their forward motions "so far as they are not retarded by the resistance of the air, or impelled downward by the force of gravity," and he referred also to "the greater bodies of planets and comets." (On the inertial aspect of the motion of "greater bodies" such as "planets and comets," see Supplement 12.) At

*The earliest known statement of this law was made by René Descartes in a book that he did not publish. It appeared in print for the first time in a work by Pierre Gassendi. But prior to Newton's *Principia* there was no completely developed inertial physics. It is not without significance that this early book of Descartes was based on the Copernican point of view; Descartes suppressed it on learning of the condemnation of Galileo. Gassendi likewise was a Copernican. He actually made experiments with objects let fall from moving ships and moving carriages to test Galileo's conclusions about inertial motion. Descartes first published his version of the law of inertia in his *Principles of Philosophy* (1644); the earlier statement, in Descartes's *The World,* was published after Descartes's death in 1650. See Suppl. 8.

this one stroke Newton postulated the opposite view of Aristotelian physics. In the latter, no celestial body could move uniformly in a straight line in the absence of a force, because this would be a "violent" motion and so contrary to its nature. Nor could a terrestrial object, as we have seen, move along its "natural" straight line without an external mover or an internal motive force. Newton, presenting a physics that applies simultaneously to both terrestrial and celestial objects, stated that in the absence of a force bodies do not necessarily stand still or come to rest as Aristotle supposed, but they may move at constant rectilinear speed. This "indifference" of all sorts of bodies to rest or uniform straight-line motion in the absence of a force clearly is an advanced form of Galileo's statement in his book on sunspots (p. 88), the difference being that in that work Galileo was writing about uniform motion along a great spherical surface concentric with the earth.

Newton said of the laws of motion that they were "such principles as have been received by mathematicians, and . . . confirmed by [an] abundance of experiments. By the first two Laws and the first two Corollaries, Galileo discovered that the descent of bodies varies as the square of the time and that the motion of projectiles is in the curve of a parabola, experience agreeing with both, unless so far as these motions are a little retarded by the resistance of the air." The "two Corollaries" deal with methods used by Galileo and many of his predecessors to combine two different forces or two independent motions. Fifty years after the publication of Galileo's *Two New Sciences* it was difficult for Newton, who had already established an inertial physics, to conceive that Galileo could have come as close as he had to the concept of inertia without having taken full leave of circularity and having known the true principle of linear inertia.

Newton was being very generous to Galileo because, however it may be argued that Galileo "really did" have the law of inertia or Newton's Law I, a great stretch of the imagination is required to assign any credit to Galileo for Law II. This law has two parts. In the second half of Newton's statement of Law II, the "change in motion" produced by an "impressed" or "motive" force—whether that is a change in the speed with which a body moves or a change in the direction in which it is moving—is said to be "in the direction of the straight line along which that force is impressed." This much is certainly implied in Galileo's analysis of projectile motion because Galileo assumed that in the forward direction there is no acceleration because there is no horizontal force, except the negligible action of air friction; but in the vertical direction there is an acceleration or continual increase of downward speed, because of the downward-acting weight force. But the first part of Law II—that the change in the magnitude of the motion is related to the motive force—is something else again; only a Newton could have seen it in Galileo's studies of falling bodies. This part of the law says that if an object were to be acted on first by one force F_1 and then by some other force F_2, the accelerations or changes in speed produced, A_1 and A_2, would be proportional to the forces, or that

$$\frac{F_1}{F_2} = \frac{A_1}{A_2}, \text{ or}$$

$$\frac{F_1}{A_1} = \frac{F_2}{A_2}$$

But in analyzing falling, Galileo was dealing with a situation in which only one force acted on each body, its weight W, and the acceleration it produced was g the acceleration of a freely falling body. (For the two forms of Newton's Law II, see p. 184.)

Where Aristotle had said that a given force gives an object a certain characteristic speed, Newton now said that a given force always produces in that body a definite acceleration A. To find the speed V, we must know how long a time T the force has acted, or how long the object has been accelerated, so that Galileo's law

$$V = AT$$

may be applied.

At this point let us try a thought-experiment, in which we assume we have two cubes of aluminum, one just twice the volume of the other. (Incidentally, to "duplicate" a cube—or make a cube having exactly twice the volume as some given cube—is as impossible within the framework of Euclidean geometry as to trisect an angle or to square a circle.) We now subject the smaller cube to a series of forces F_1, F_2, F_3, . . . and determine the corresponding accelerations A_1, A_2, A_3, . . . In accordance with Law II, we would find that there is a certain constant value of the ratio of force to acceleration

$$\frac{F_1}{A_1} = \frac{F_2}{A_2} = \frac{F_3}{A_3} = \ldots\ldots = m_s$$

which for this object we may call m_s. We now repeat the operations with the larger cube and find that the same set of forces F_1, F_2, F_3, . . . respectively produces *another* set of accelerations a_1, a_2, a_3, . . . In accordance with Newton's second law, the force-acceleration ratio is again a constant which for this object we may call m_l

$$\frac{F_1}{a_1} = \frac{F_2}{a_2} = \frac{F_3}{a_3} = \ldots\ldots = m_l$$

For the larger object the constant proves to be just twice as large as the constant obtained for the smaller one and, in general, so long as we deal with a single variety of matter like pure aluminum, *this constant* is proportional to the volume and so *is a measure of the amount of aluminum in any sample.* This particular constant is a measure of an object's resistance to acceleration, or a measure of the tendency of that object to stay as it is—either at rest, or in motion in a straight line. For observe that m_l was twice m_s; to give both objects the same acceleration or change in motion the force required for the larger object is just twice what it must be for the smaller. The tendency of any object to continue in its state of motion (at constant speed in a straight line) or its state of rest is called its *inertia;* hence, Newton's Law I is also called the principle of inertia. The constant determined by finding the constant force-acceleration ratio for any given body may thus be called the body's inertia. But for our aluminum blocks this same constant is also a measure of the "quantity of matter" in the object, which is called its *mass.* We now make precise the condition that two objects of different material—say one of brass and the other of wood—shall have the same "quantity of matter": it is that they have the *same mass* as determined by the force-acceleration ratio, or the *same inertia.*

In ordinary life, we do not compare the "quantity of matter" in objects in terms of their inertias, but in terms of their weight. Newtonian physics makes it clear why we can, and through its clarification we are able to understand why at any place on the earth two unequal weights in a vacuum fall at the same rate. But we may observe that in at least one common situation we always compare the inertias of objects rather than their weights. This happens when a person hefts two objects to find which is heavier, or has the greater mass. He does not hold them out to see which pulls down more on his arm; instead, he moves them up and down to find which is easier to move. In this way he determines which has the greater resistance to a change in its state of motion in a straight line or of rest—that is, which has the greater inertia. (On Newton's concept of inertia, see Supplement 15.)

FINAL FORMULATION OF THE LAW OF INERTIA

At one point in his *Discourses and Demonstrations Concerning Two New Sciences,* Galileo imagined a ball to be rolling along a plane and noted that "equable motion on this plane would be perpetual if the plane were of infinite extent." A plane without limit is all right for a pure mathematician, who is a Platonist in any case. But Galileo was a man who combined just such a Platonism with a concern for applications to the real world of sensory experience. In the *Two New Sciences,* Galileo was not interested only in abstractions as such, but in the analysis of real motions on or near the earth. So we understand that having talked about a plane without limit, he does not continue with such a fancy, but asks what would happen on such a plane if it were a real earthly plane, which for him means that it is "ended, and [situated] on high." The ball, in the real world of physics, falls off the plane and begins to fall to the ground. In this case,

> the movable (which I conceive of as being endowed with heaviness), driven to the end of this plane and going on further, adds on to its

> previous equable and indelible motion that downward tendency which it has from its own heaviness. Thus there emerges a certain motion, compounded from equable horizontal and from naturally accelerated downward [motion], which I call "projection."

Unlike Galileo, Newton made a clear separation between the world of abstract mathematics and the world of physics, which he still called philosophy. Thus the *Principia* included both "mathematical principles" as such and those that could be applied in "natural philosophy," but Galileo's *Two New Sciences* included only those mathematical conditions exemplified in nature. For instance, Newton plainly knew that the attractive force exerted by the sun on a planet varies as the inverse-square of the distance

$$F \propto \frac{1}{D^2}$$

but in Book One of the *Principia* he explored the consequences not only of this particular force but of others with quite different dependence on the distance, including

$$F \propto D$$

"THE SYSTEM OF THE WORLD"

At the beginning of Book Three, which was devoted to "The System of the World," Newton explained how it differed from the preceding two, which had been dealing with "The Motion of Bodies":

> In the preceding Books I have laid down . . . principles not philosophical [pertaining to physics] but mathematical: such, namely, as we may build our reasonings upon in philosophical inquiries. These principles are laws and conditions of certain motions, and powers or forces, which chiefly have respect to philosophy; but, lest they should have appeared of themselves dry and barren, I have illustrated them here and there with some philosophical scholiums, giving an account of such things as are of a more general nature, and which philosophy seems chiefly to be founded on: such as the density and the resistance of bodies, spaces void of all bodies, and the motion of light and

sounds. It remains that, from the same principles, I now demonstrate the structure of the System of the World.

I believe it fair to say that it was the freedom to consider problems either in a purely mathematical way or in a "philosophical" (or physical) way that enabled Newton to express the first law and to develop a complete inertial physics. After all, physics as a science may be developed in a mathematical way but it always must rest on experience—and experience never shows us pure inertial motion. Even in the limited examples of linear inertia discussed by Galileo, there was always some air friction and the motion ceased almost at once, as when a projectile strikes the ground. In the whole range of physics explored by Galileo there is no example of a physical object that has even a component of pure inertial motion for more than a very short time. It was perhaps for this reason that Galileo never framed a general law of inertia. He was too much a physicist.

But as a mathematician Newton could easily conceive of a body's moving along a straight line at constant speed forever. The concept "forever," which implies an infinite universe, held no terror for him. Observe that his statement of the law of inertia, that it is the natural condition for bodies to move in straight lines at a constant speed, occurs in Book One of the *Principia,* the portion said by him to be mathematical rather than physical. Now, if it is the natural condition of motion for bodies to move uniformly in straight lines, then this kind of inertial motion must characterize the planets. The planets, however, do not move in straight lines, but rather along ellipses. Using a kind of Galilean approach to this single problem, Newton could say that the planets must therefore be subject to two motions: one inertial (along a straight line at constant speed) and one always at right angles to that straight line drawing each planet toward its orbit. (See, further, Supplements 11 and 12.)

Though not moving in a straight line, each planet nevertheless represents the best example of inertial motion observable in the universe. Were it not for that component of inertial motion, the force that continually draws the planet away from the straight line would draw the planet in toward the sun until the two bodies collided. Newton once used this argument to prove the existence of God. If the planets had not received a push to give them an inertial (or tangential) component of motion, he said, the solar attractive force would not draw them into an orbit but instead would move each planet in a straight line toward the sun itself. Hence the universe could not be explained in terms of matter alone.

For Galileo pure circular motion could still be inertial, as in the example of an object on or near the surface of the earth. But for Newton pure circular motion was not inertial; it was accelerated and required a force for its continuance. Thus it was Newton who finally shattered the bonds of "circularity" which still had held Galileo in thrall. And so we may understand that it was Newton who showed how to build a celestial mechanics based on the laws of motion, since the elliptical (or almost circular) orbital motion of planets is not purely inertial, but requires additionally the constant action of a force, which turns out to be the force of universal gravitation.

Thus Newton, again unlike Galileo, set out to "demonstrate the structure of the System of the World," or—as we would say today—to show how the general laws of terrestrial motion may be applied to the planets and to their satellites.

In the first theorem of the *Principia* Newton showed that if a body were to move with a purely inertial motion, then with respect to any point not on the line of motion, the law of equal areas must apply. In other words, a line drawn from any such body to such a point will sweep out equal areas in equal times. Conceive a body moving with purely inertial motion along the straight line of which *PQ* is a segment. Then in a set of equal time intervals (Fig. 30) the body will move through equal distances *AB, BC, CD,* . . . because, as Galileo showed, in uniform motion a body moves through equal distances in equal times. But observe that a line from the point *O* sweeps out equal areas in these equal times, or that the areas of triangles *OAB, OBC, OCD,* . . . are equal. The reason is that the area of a triangle is one-half the product of its altitude and its base; and all these triangles have the same altitude *OH* and equal bases. Since

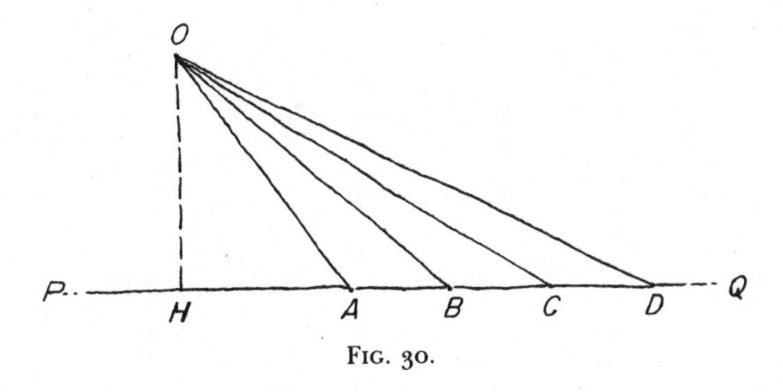

FIG. 30.

$$AB = BC = CD = \ldots$$

it is true that

$$\tfrac{1}{2}AB \times OH = \tfrac{1}{2}BC \times OH = \tfrac{1}{2}CD \times OH = \ldots\ldots$$

or

$$\text{area of } \Delta OAB = \text{area of } \Delta OBC = \text{area of } \Delta OCD = \ldots$$

Thus the very first theorem proved in the *Principia* showed that purely inertial motion leads to a law of equal areas, and so is related to Kepler's second law. Newton then proved that if at regular intervals of time, a body moving with purely inertial motion were to receive a momentary impulse (a force acting for an instant only), all these impulses being directed toward the same point *S*, then the body would move in each of the equal time-intervals between impulses so that a line from it to *S* would sweep out equal areas. This situation is shown in Fig. 31. When the body reaches the point *B* it receives an impulse toward *S*. The new motion is a combination of the original motion along *AB* and a motion toward *S*, which produces a uniform rectilinear motion toward *C*, etc.: The triangles *SAB*, *SBC*, and *SCD* . . . have the same area. The next step, according to Newton, is as follows:

> . . . Now let the number of those triangles be augmented, and their breadth diminished *in infinitum;* and (by Cor. iv, Lem. iii) their ultimate perimeter *ADF* will be a curved line: and therefore the centripetal

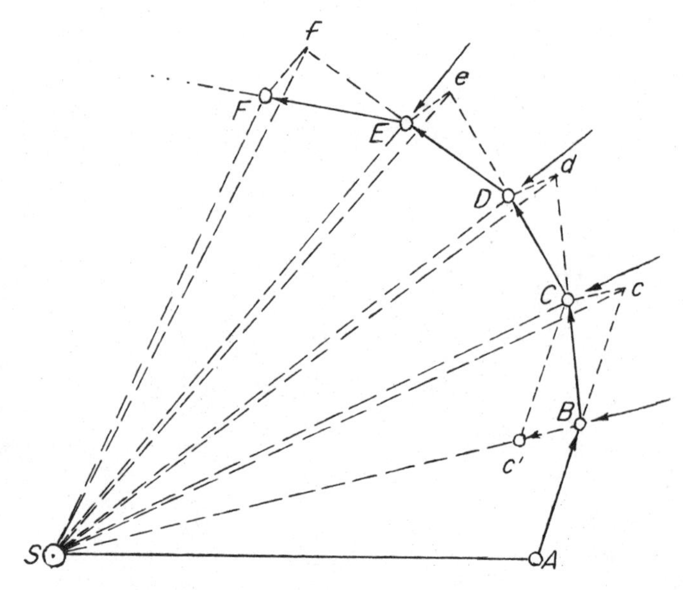

FIG. 31. If at *B* the body had received no impulse, it would, during time *T*, have moved along the continuation of *AB* to *c*. The impulse at *B*, however, gives the body a component of motion toward *S*. During *T* if the body's only motion came from that impulse, it would have moved from *B* to *c'*. The combination of these two movements, *Bc* and *Bc'*, results during time *T* in a movement from *B* to *C*. Newton proved that the area of the triangle *SBC* is equal to the area of the triangle *SBc*. Hence, even when there is an impulsive force directed toward *S*, the law of equal areas holds.

force, by which the body is continually drawn back from the tangent of this curve, will act continually; and any described areas *SADS*, *SAFS*, which are always proportional to the times of description, will, in this case also, be proportional to those times. Q.E.D.

In this way Newton proceeded to prove:

Proposition 1. Theorem 1.

The areas which revolving bodies describe by radii drawn to an immovable centre of force do lie in the same immovable planes, and are proportional to the times in which they are described.

In simple language, Newton proved in the first theorem of Book One of the *Principia* that if a body is continually drawn toward some center of force, its otherwise inertial motion will be transformed into motion along a curve, and that a line from the center of force to the body will sweep out equal areas in equal times. In proposition 2 (theorem 2) he proved that if a body moves along a curve so that the areas described by a line from the body to any point are proportional to the times, there must be a "central" (centripetal) force continuously urging the body toward that point. The significance of Kepler's law of elliptical orbits does not appear until proposition 11: to find "the law of the centripetal force tending to a focus of the ellipse." This force varies "inversely as the square of the distance." Then Newton proves that if a body moving in an hyperbola or in a parabola is acted on by a centripetal force tending to a focus, the force still varies inversely as the square of the distance. Several theorems later, in proposition 17, Newton proves the converse, that if a body moves subject to a centripetal force varying inversely as the square of the distance, the path of the body must be a conic section: an ellipse, a parabola, or a hyperbola. (See Supplement 13).

We may note that Newton has treated Kepler's laws exactly in the same order as Kepler himself: first the law of areas as a general theorem, and only later the particular shape of planetary orbits as ellipses. What seemed at first to be a rather odd way of proceeding has been shown to represent a fundamental logical progression of a kind that is the opposite of the sequence that would have been followed in an empirical or observational approach.

In Newton's reasoning about the action of a centripetal force on a body moving with purely inertial motion, mathematical analysis, for the first time, disclosed the true meaning of Kepler's second law, of equal areas! Newton's reasoning showed that this law implies a center of force for the motion of each planet. Since the equal areas in planetary motion are reckoned with respect to the sun, Kepler's second law becomes in Newton's treatment the basis for proving geometrically that a central force emanating from the sun attracts all the planets.

So much for the problem raised by Halley. Had Newton stopped his work at this point, we would still admire his achievement enormously. But Newton went on, and the results were even more outstanding.

THE MASTERSTROKE: UNIVERSAL GRAVITATION

In Book Three of the *Principia,* Newton showed that as Jupiter's satellites move in orbits around their planet, a line from Jupiter to each satellite will "describe areas proportional to the times of description," and that the ratio of the squares of their times to the cubes of their mean distances from the center of Jupiter is a constant, although a constant having a different value from the constant for the motion of the planets. Thus if T_1, T_2, T_3, T_4 be the periodic times of the satellites, and a_1, a_2, a_3, a_4 be their respective mean distances from Jupiter,

$$\frac{(a_1)^3}{(T_1)^2} = \frac{(a_2)^3}{(T_3)^2} = \frac{(a_3)^3}{(T_3)^2} = \frac{(a_4)^3}{(T_4)^2}$$

Not only do these laws of Kepler apply to the Jovian system, but they also apply to the five satellites of Saturn known to Newton —a result wholly unknown to Kepler. The third law of Kepler could not be applied to the earth's moon because there is only one moon, but Newton did state that its motion agrees with the law of equal areas. Hence, one may see that there is a central force, varying as the inverse-square of the distance, that holds each planet to an orbit around the sun and each planetary satellite to an orbit around its planet.

Now Newton makes the masterstroke. He shows that a single universal force (a) keeps the planets in their orbits around the sun, (b) holds the satellites in their orbits, (c) causes falling objects to descend as observed, (d) holds objects on the earth, and (e) causes the tides. It is the force called *universal gravity,* and its fundamental law may be written

$$F = G\frac{mm'}{D^2}$$

This law says that between any two bodies whatsoever, of masses m and m', wherever they may be in the universe, separated by a distance D, there is a force of attraction that is *mutual*, and each body attracts the other with a force of identical magnitude, which is *directly proportional to the product of the two masses* and *inversely proportional to the square of the distance between them.* G is a constant of proportionality, and it has the same value in all circumstances —whether in the mutual attraction of a stone and the earth, of the earth and the moon, of the sun and Jupiter, of one star and another, or of two pebbles on a beach. This constant G is called the *constant of universal gravitation* and may be compared to other "universal" constants—of which there are not very many in the whole of science—such as c, the speed of light, which figures so prominently in relativity, or h, Planck's constant, which is so basic in quantum theory.

How did Newton find his law? It is difficult to tell in detail, but we can reconstruct some of the basic aspects of the discovery.

From a later memorandum (about 1714), we learn that Newton as a young man "began to think of gravity extending to the orb of the moon, and having found out how to estimate the force with which [a] globe revolving within a sphere presses the surface of the sphere, from Kepler's rule of the periodical times of the planets being in a sesquialterate proportion [i.e., as the 3/2 power] of their distances from the centers of their orbs, I deduced that the forces which keep the planets in their orbs must [be] reciprocally as the squares of their distances from the centers about which they revolve: and thereby compared the force requisite to keep the moon in her orb with the force of gravity at the surface of the earth, and found them answer [i.e., agree] pretty nearly."

With this statement as guide, let us consider first a globe of mass m and speed v moving along a circle of radius r. Then, as Newton found out, and as the great Dutch physicist Christiaan Huygens (1629–1695) also discovered (and to Newton's chagrin, published first; see Supplement 14), there must be a central acceleration, of magnitude v^2/r. That is, an acceleration follows from the fact that the globe is not at rest nor moving at constant speed in a straight line; from Law I and Law II, there must be a force and hence an acceleration. We shall not prove that this acceleration has a magnitude v^2/r, but that it is directed toward the center you can see if you whirl a ball in a circle at the end of a string. A force is needed to pull the ball constantly toward the center, and from Law II the acceleration must always have the same direction as the accelerating force. Thus for a planet of mass m, moving approximately in a circle of radius r at speed v, there must be a central force F of magnitude

$$F = mA = m\frac{v^2}{r}.$$

If T is the period, or time for the planet to move through 360°, then in time T the planet moves once around a circle of radius r, or through a circumference of $2\pi r$. Hence the speed v is $2\pi r/T$, and

$$F = mA = mv^2 \times \frac{1}{r} = m\left[\frac{2\pi r}{T}\right]^2 \times \frac{1}{r}$$

$$= m \times \frac{4\pi^2 r^2}{T^2} \times \frac{1}{r}$$

$$= m \times \frac{4\pi^2 r^2}{T^2} \times \frac{1}{r} \times \frac{r}{r}$$

$$= \frac{4\pi^2 m \times r^3}{T^2 \times r^2} = \frac{4\pi^2 m}{r^2} \times \frac{r^3}{T^2}.$$

Since for every planet in the solar system, r^3/T^2 has the same value K (by Kepler's rule or third law),

$$F = \frac{4\pi^2 m}{r^2} \times K = 4\pi^2 K \frac{m}{r^2}.$$

The radius r of the circular orbit corresponds in reality to D the average distance of a planet from the sun. Hence, for any planet the law of force keeping it in its orbit must be

$$F = 4\pi^2 K \frac{m}{D^2}$$

where m is the mass of the planet, D is the average distance of the planet from the sun, K is "Kepler's constant" for the solar system (equal to the cube of the mean distance of any planet from the sun divided by the square of its period of revolution), and F is the force with which the sun attracts the planet and draws it continually off its purely inertial path into an ellipse. Thus far mathematics and logic may lead a man of superior wit who knows the Newtonian laws of motion and the principles of circular motion.

But now we rewrite the equation as

$$F = \left[\frac{4\pi^2 K}{M_s}\right] \frac{M_s m}{D^2}$$

where M_s is the mass of the sun and say that the quantity

$$\frac{4\pi^2 K}{M_s} = G$$

is a *universal constant*, that the law

$$F = G \frac{M_s m}{D^2}$$

is not limited to the force between the sun and a planet. It applies also to every pair of objects in the universe, M_s and m becoming the masses m and m' of those two objects and D becoming the distance between them:

$$F = G \frac{mm'}{D^2}$$

There is no mathematics—whether algebra, geometry, or the calculus—to justify this bold step. One can say of it only that it is one of those triumphs that humble ordinary men in the presence of genius. And just think what this law implies. For instance, this book that you hold in your hands attracts the sun in a calculable degree; the same force makes the moon follow its orbit and an apple fall from the tree. Late in life Newton said it was this last comparison that inspired his great discovery. (See, further, Supplement 14.)

The moon (see Fig. 32) if not attracted by the earth would have a purely inertial motion and in a small time t would move uniformly along a straight line (a tangent) from A to B. It does not, said Newton, because while its inertial motion would have carried it from A to B, the gravitational attraction of the earth will have made it fall toward the earth from the line AB to C. Thus the moon's departure from a purely inertial rectilinear path is caused by its continual "falling" toward the earth—and its falling is just like the falling of an apple. Is this true? Well, Newton put the proposition to a test, as follows:

Why does an apple of mass m fall to the earth? It does so, we may now say, because there is a force of universal gravitation between it and the earth, whose mass is M_e. But what is the distance between the earth and the apple? Is it the few feet from the apple to the ground? The answer to this question is far from obvious. Newton eventually was able to prove that the attraction

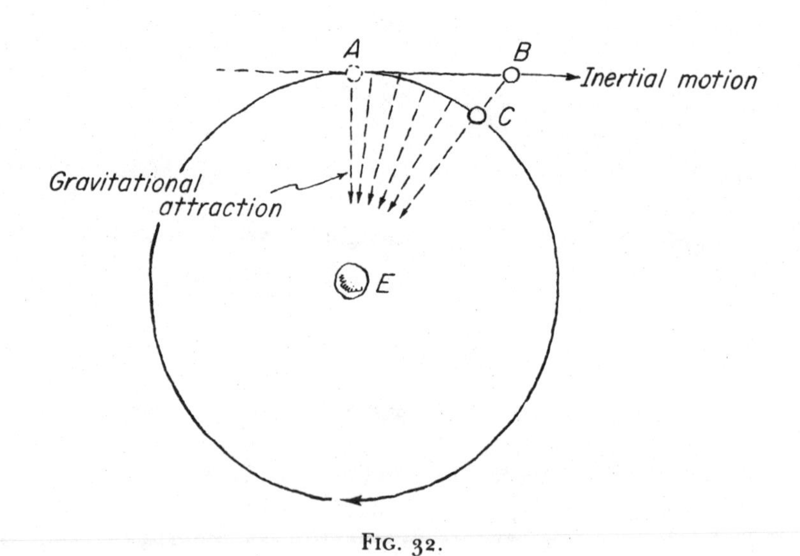

FIG. 32.

between a small object and a more or less homogeneous and more or less spherical body is exactly the same as if all the large mass of the body were concentrated at its geometric center. This theorem means that in considering the mutual attraction of earth and apple, the distance D in the law of universal gravitation may be taken to be the earth's radius, R_e. Hence the law states that the attraction between the earth and an apple is:

$$F = G\frac{mM_e}{R_e^{\ 2}},$$

where m is the mass of the apple, M_e the mass of the earth, and R_e the earth's radius. But this is an expression for the *weight* W of the apple, because the weight of any terrestrial object is merely the magnitude of the force with which it is gravitationally attracted by the earth. Thus,

$$W = G\frac{mM_e}{R_e^{\ 2}}.$$

There is a second way of writing an equation for the weight of an apple or of any other terrestrial object of mass m. We use Newton's Law II, which says that the mass m of any object is the ratio of the force acting on the object to the acceleration produced by that force,

$$m = \frac{F}{A}$$

or

$$F = mA.$$

Note that when an apple falls from the tree, the force pulling it down is its weight W, so that

$$W = mA.$$

Since we now have two different mathematical statements of the same force or weight W, they must be equal to each other, or

$$mA = G\frac{mM_e}{R_e{}^2}$$

and we can divide both sides by *m* to get

$$A = G\frac{M_e}{R_e{}^2}.$$

So, by Newtonian principles, we have at once explained why at any spot on this earth all objects—whatever their mass *m* or weight *W* may be—will have the same acceleration *A* when they fall freely, as in a vacuum. The last equation shows that this acceleration of free fall is determined by the mass M_e and radius R_e of the earth and a universal constant *G*, none of which *depends in any way* on the particular mass *m* or weight *W* of the falling body.

Now let us write the last equation in a slightly different way,

$$A = G\frac{M_e}{D_e{}^2}$$

where D_e stands for the distance from the center of the earth. At or near the earth's surface D_e is merely the earth's radius R_e. Now consider a body placed at a distance D_e of 60 earth-radii from the earth's center. With what acceleration A' will it fall toward the center of the earth? The acceleration A' will be

$$A' = G\frac{M_e}{(60\ R_e)^2} = G\frac{M_e}{3600\ R_e{}^2} = \frac{1}{3600}\ G\frac{M_e}{R_e{}^2}.$$

We just saw that at the surface of the earth an apple or any other object will have a downward acceleration equal to $G\frac{M_e}{R_e{}^2}$, and now we have proved that an object at 60 earth-radii will have an acceleration just 1/3600th of that value. On the average, a body at the earth's surface falls in one second toward the earth through a distance of 16.08 feet, so that out at a distance of 60 earth-radii from the earth's center a body should fall

$$1/3600 \times 16.08 \text{ feet} = 1/3600 \times 16.08 \times 12 \text{ inches} =$$

$$0.0536 \text{ inches.}$$

It happens that there is a body, our moon, out in space at a distance of 60 earth-radii and so Newton had an object for testing his theory of universal gravitation. If the same gravitational force makes both the apple and the moon fall, then in one second the moon should fall through 0.0536 inches from its inertial path to stay on its orbit. A rough computation, based on the simplifying assumptions that the moon's orbit is a perfect circle and that the moon moves uniformly without being affected by the gravitational attraction of the sun, yields a distance fallen in one second of 0.0539 inches—or a remarkable agreement to within 0.0003 inches! Another way of seeing how closely observation agrees with theory is to observe that the two values differ by 3 parts in about 500, which is the same as 6 parts in 1000 or 0.6 parts per hundred (0.6 per cent). Another way of seeing how this calculation can be made (perhaps following the lead Newton himself gave in the quotation on page 165) is as follows:

1) For a body on earth (the apple), the acceleration (g) of free fall is

$$g = G\frac{M_e}{R_e^{\,2}}.$$

2) For the moon, the form of Kepler's third law is

$$k = \frac{R_m^{\,3}}{T_m^{\,2}}$$

where R_m and T_m are respectively the radius of the moon's orbit and the moon's period of revolution. If the gravitational force *is* universal, then the relation derived earlier for planets moving around the sun

$$G = \frac{4\pi^2 K}{M_s}$$

can be rewritten for the moon moving around the earth, in the form

$$G = \frac{4\pi^2 k}{M_e}.$$

Hence, we may compute g from Equation (1), as follows:

$$g = \left[\frac{4\pi^2 k}{M_e}\right] \frac{M_e}{R_e^{\ 2}} = 4\pi^2 k \left[\frac{1}{R_e^{\ 2}}\right]$$

$$= 4\pi^2 \left[\frac{R_m^{\ 3}}{\mathrm{T_m}^{\ 2}}\right]\left[\frac{1}{R_e^{\ 2}}\right] = 4\pi^2 \left[\frac{R_m^{\ 3}}{T_m^{\ 2}}\right]\left[\frac{1}{R_e^{\ 2}}\right]\left[\frac{R_e}{R_e}\right]$$

$$= 4\pi^2 \left[\frac{R_m}{R_e}\right]^3 \left[\frac{R_e}{T_m^{\ 2}}\right].$$

Since

$$\frac{R_m}{R_e} = 60, \text{ and } R_e = 4{,}000 \times 5{,}280 \text{ feet}$$

$$T_m = 28d = 28 \times 24 \times 3600 \text{ sec}$$

we may compute that

$$g \approx 32 \text{ ft/sec}^2$$

or

$$g \approx 1000 \text{ cm/sec}^2.$$

Newton said, in the autobiographical memorandum I have quoted, that he "compared the force requisite to keep the moon in her orb with the force of gravity at the surface of the earth."

In Book Three of the *Principia,* Newton shows that the moon, in order to keep along its observed orbit, falls away from its straight line inertial path through a distance of 15 1/12 Paris feet (an old measure) in every minute. Imagine the moon, he says, "deprived of all motion to be let go, so as to descend toward the earth with the impulse of all that force by which . . . it is retained in its orb." In one minute of time it will descend through the same distance that it does when this descent occurs together with the normal inertial motion. Next, assume that this motion toward the earth is due to gravity, a force that varies inversely as the square of the distance. Then, at the surface of the earth this force would be greater by a factor 60 × 60 than at the moon's orbit. Since the acceleration is, by Newton's second law, proportional to the accelerating force, a body brought from the moon's orbit to the earth's surface would have an increase in its acceleration of 60 × 60. Thus, Newton argues, if gravity is a force varying inversely as the square of the distance, a body at the earth's surface should fall, starting from rest, through a distance of nearly 60 × 60 × 15 1/12 Paris feet in one minute, or 15 1/12 Paris feet in one second.

From Huygens's pendulum experiment Newton obtained the result that on earth (at the latitude of Paris) a body falls just about that far. Thus he proved that it is the force of the earth's gravity that retains the moon in its orbit. In making the computation, Newton predicted from observations of the moon's motion and from gravitation theory that the distance fallen by a body on earth in one second would be 15 Paris feet, 1 inch and 1 4/9 lines (1 line = 1/12 inch). Huygens's result for free fall at Paris was 15 Paris feet, 1 inch, 1 7/9 lines. The difference was 3/9 or 1/3 of a line and hence 1/36 of an inch—a very small number indeed. By the time Newton wrote the *Principia,* he had found a far better agreement between theory and observation than in that rough test he had made twenty years earlier.

Newton said that in this test observation agreed with prediction "pretty nearly." Two factors were involved in that phrase. First, he chose a poor value of the earth's radius and so obtained bad numerical results, agreeing only roughly or "pretty nearly." Second, since he had not then been able to prove rigorously that a homogeneous sphere attracts gravitationally as if all its mass were concentrated at its center, the proof was at best rough and approximate.

But this test proved to Newton that his concept of universal gravitation was valid. You can appreciate how remarkable it was when you consider the nature of the constant G. We saw earlier that $G = \frac{4\pi^2 K}{M_s}$ and we may well ask what either K (the cube of any planet's distance from the sun divided by the square of the periodic time of that planet's revolution about the sun) or M_s (the mass of the sun) has to do with either the earth's pull on a stone or the earth's pull on the moon. If the fact that the earth happens to be within the solar system lessens the wonder that G should apply to the stone and the moon, consider a system of double stars located millions of light-years away from the solar system. Such a pair of stars may form an eclipsing binary, in which one of the stars encircles the other as the moon encircles the earth. Way out there, beyond any possible influence of the sun, the same constant $G = \frac{4\pi^2 K}{M_s}$ applies to the attraction of each of the stars by the other. This is a universal constant *in spite of the fact* that in the form in which Newton discovered it, it was based on elements in *our solar system.* Evidently, the act of dividing the Kepler constant by the mass of the central body about which the others revolve eliminates any special aspects of that particular system—whether of planets revolving about the sun, or satellites revolving about Jupiter or Saturn. (See, further, Supplement 15.)

THE DIMENSIONS OF THE ACHIEVEMENT

A few further achievements of Newtonian dynamics, or gravitation theory, will enable us to comprehend its heroic dimensions. Suppose the earth were not quite a perfect sphere, but were oblate—flattened at the poles and bulging at the equator. Consider now the acceleration A of a freely falling body at a pole, at the equator, and at two intermediate points a and b. Clearly the "radius" R of the earth, or distance from the center, would increase from the pole to the equator, so that

$$R_p < R_b < R_a < R_e.$$

As a result the acceleration A of free fall at those places would have different values:

$$A_p = G\frac{M_e}{R_p^{\ 2}};\ A_b = G\frac{M_e}{R_b^{\ 2}};\ A_a = \frac{M_e}{R_a^{\ 2}};\ A_e = \frac{M_e}{R_e^{\ 2}},$$

so that

$$A_p > A_b > A_a > A_e.$$

The following data, obtained from actual experiment, show the acceleration varies with latitude:

Latitude	*Acceleration of free fall*	
0° (equator)	978.039 cm/sec^2	32.0878 ft/sec^2
20°	978.641	32.1076
40°	980.171	32.1578
60°	981.918	32.2151
90°	983.217	32.2577

In Newton's day, the acceleration of free fall was found by determining the length of a seconds pendulum—one that has a period of 2 seconds. The equation for the period T of a simple pendulum swinging through a short arc is

$$T = 2\pi\sqrt{\frac{l}{g}}$$

where l is the length of the pendulum (computed from the point of support to the center of the bob) and g is the acceleration of free fall. Halley found that when he went from London to St. Helena it was necessary to shorten the length of his pendulum in order to have it continue to beat seconds. Newton's mechanics not only explains this variation but leads to a prediction of the shape of the earth, an oblate spheroid, flattened at the poles and bulging at the equator.

The variations in g, the acceleration of free fall, imply concomitant variations in the weight of any physical object transported from one latitude to another. A complete analysis of this variation in weight requires the consideration of a second factor, the force arising from the rotation of the object along with the earth. The factor that enters here is v^2/r where v is the linear speed along a circle and r the circle's radius. At different latitudes, there will be different values of both v and r. Furthermore, to relate the rotational effect to weight, a component must be taken along a line from the center of the earth to the position in question, since the rotational effect occurs in the plane of circular motion, or along a parallel of latitude. It is because of these rotational forces that the earth, according to Newtonian physics, acquired its shape.

A second consequence of the equatorial bulge is the precession of the equinoxes. In actual fact, the difference between the polar and equatorial radii of the earth may not seem very great:

equatorial radius = 6378.388 km = 3963.44 miles
polar radius = 6356.909 km = 3949.99 miles

But if we represent the earth with an 18-inch globe, the difference between the smallest and greatest diameters would be about 1/16th of an inch. Newton showed that precession occurs because the earth is spinning on an axis inclined to the plane of its orbit, the plane of the ecliptic. In addition to the gravitational attraction that keeps the earth in its orbit, the sun exerts a pull on the bulge, thus tending to straighten the axis. This force of the sun tends to make the earth's axis perpendicular to the plane of the ecliptic (Fig. 33A) or make the plane of the bulge (or of the earth's equator) coincide with the plane of the ecliptic. At the same time the moon's pull tends to make the plane of the bulge coincide with the plane of its orbit (inclined at about 5 degrees to the plane of the ecliptic). The moon's force is somewhat greater in this regard than the sun's. If the earth were a perfect sphere, the pull on it by the sun or moon would be symmetrical and there would be no tendency for the axis of rotation to "straighten out"; the lines of action of the gravitational pulls of sun and moon would pass through the earth's center. But if the earth should be oblate, or flattened at the poles, as Newton supposed, then there would be a net force tending to shift the

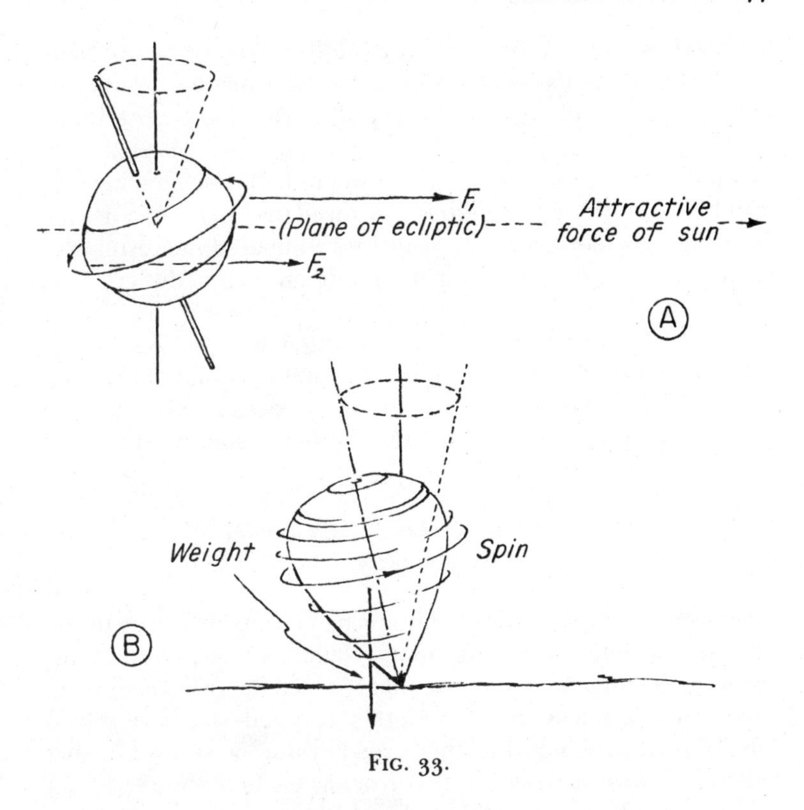

FIG. 33.

axis of the earth. And accordingly there would be a predictable effect.

Now it is a result in Newtonian physics that if a force is exerted so as to change the orientation of the axis of a spinning body, the effect will be that the axis itself, rather than changing its orientation, will undergo a conical motion. This effect may be seen in a spinning top. The axis of rotation is usually not absolutely vertical. The weight of the top acts therefore to turn the axis about the spinning point so as to make the axis horizontal. The weight tends to produce a rotation whose axis is at right angles to that of the top's spin, and the result is the conical motion of the axis shown in Fig. 33B. The phenomenon of precession had been known since its discovery in the second century B.C. by

Hipparchus, but its cause had been wholly unknown before Newton. Newton's explanation not only resolved an ancient mystery, but was an example of how one could predict the precise shape of the earth by applying theory to astronomical observations. Newton's predictions were verified when the French mathematician Pierre L. M. de Maupertuis measured the length of a degree of arc along a meridian in Lapland and compared the result with the length of a degree along the meridian nearer the equator. The result was an impressive victory for the new science.

Yet another achievement of the Newtonian theory was a general explanation of the tides relating them to gravitational action of the sun and moon on the waters of the oceans. We may well understand the spirit of admiration that inspired Alexander Pope's famous couplet:

> *Nature, and Nature's Laws lay hid in Night.*
> *God said,* Let Newton be! *and All was* Light.

In seeing how the Newtonian mechanics enabled man to explain the motions of planets, moons, falling stones, tides, trains, automobiles, and anything else that is accelerated—speeded up, slowed down, started in its motion or stopped—we have solved our original problem. But there remain one or two items that require a word or so more. It is true, as Galileo observed, that for ordinary bodies on the earth (which may be considered as revolving in a large elliptic orbit at an average distance from the sun of about 93 million miles), the situation is very much like being on something that is moving in a straight line, and there *is* an indifference to uniform rectilinear motion and to rest so far as all the dynamical problems are concerned. On the rotating earth, where the arc during any time interval, such as the flight of a bullet, is a part of a "circle" smaller than the annual orbit, another Newtonian kind of principle can be invoked, the principle of the conservation of angular momentum.

The angular momentum of a small object rotating in a circle (as a stone held on the top of a tower on a rotating earth) is given by the expression mvr where r is the radius of rotation, m the mass, and v the speed along the circle. The principle says that under a large variety of conditions (specifically, in all circumstances in which there is no external force of a special kind), the angular momentum remains constant.

An example may be given. A man stands on a whirling platform, with his arms outstretched and clutching a 10-pound weight in each hand. He is set whirling slowly on the turntable and then is told to bring his hands in toward his body along a horizontal plane so that he looks like Fig. 34. He finds that he spins faster and faster. Stretching his arms out once again will slow him down. For anyone who has never seen such a demonstration before (it is a standard figure in ice skating) the first encounter can be quite startling. Now let us see why these changes occurred. The speed v with which the masses m held in his hands move around is

$$v = \frac{2\pi r}{t}$$

where t is the time for a complete rotation, during which each mass m moves through a circumference of a circle of radius r. At first the angular momentum is

$$mvr = m \times \frac{2\pi r}{t} \times r = \frac{2\pi m r^2}{t}.$$

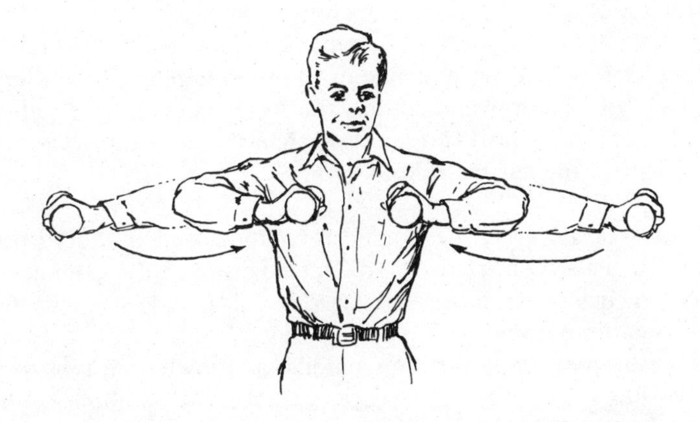

But as the man brings his arms in to his chest he makes r very much smaller. If $\frac{2\pi mr^2}{t}$ is to keep the same value, as the law of conservation demands, then t must get smaller too, which means that the time for a revolution becomes smaller as r diminishes.

What has this to do with a stone falling from a tower? At the top of the tower the radius of rotation is $R + r$ where R is the radius of the earth and r the height of the tower. When the stone strikes the ground, the radius of rotation is R. Therefore, like the masses drawn inward by the whirling mass, the stone must be moving around in a smaller circle when at the base of the tower than at the top, and so will whirl more quickly. Far from being left behind, the stone, according to our theory, should get a little ahead of the tower. How great an effect is this? Since the problem depends on t, the time for a rotation through 360 degrees, we can get a much better idea of the magnitude of the problem if we study the angular speed than if we consider some linear speed (as we did in Chapter 1). Look at the moving hands of a clock, paying particular attention to the hour hand. By how much does it appear to shift in, say, five minutes, which corresponds to dropping a ball from a much greater height than the Empire State Building? Not by any discernible degree. Now the rotation of the earth through 360 degrees takes just twice as much time as a complete rotation of the hour hand (12 hours). Since in five minutes the angular motion or rotation of the hour hand is not discernible to the unaided eye, a motion that is twice as slow produces practically no effect. Except in problems of long-range artillery firing, analysis of the movements of the trade winds, and other phenomena on a vastly larger scale than the fall of a stone, we may neglect the earth's rotation.

Such was the great Newtonian revolution, which altered the whole structure of science and, indeed, turned the course of Western civilization. How has it fared in the last 300 years? Is the Newtonian mechanics still true?

All too often the misleading statement is made that relativity theory has shown classical dynamics to be false. Nothing could be further from the truth! Relativistic corrections apply to objects moving at speeds v for which the ratio v/c is a significant quantity, c being the speed of light, or 186,000 miles per second. At the speeds attained in linear accelerators, cyclotrons, and other devices for studying atomic and subatomic particles, it is no longer true that the mass m of a physical object remains constant. Rather, it is found that the mass in motion is given by the equation

$$m = \frac{m_0}{\sqrt{1 - v^2/c^2}}$$

where m is the mass of an object moving at a speed v relative to the observer and m_0 is the mass of that same object observed at rest. Along with this revision goes Albert Einstein's now familiar equation relating mass and energy, $E = mc^2$, and the denial of the validity of Newton's belief in an "absolute" space and an "absolute" time. Well, then, might we agree with the new couplet added by J. C. Squire to the one of Pope's we have quoted?

It did not last: the Devil howling 'Ho,
Let Einstein be,' restored the status quo.

But for the whole range of problems discussed by Newton—exemplified today in the motion of stars, planets, moons, airplanes, spaceships, artificial satellites, automobiles, baseballs, rockets, and every other type of gross body—the speeds v attainable are such that v/c has to all intents and purposes the value zero and we can still apply Newtonian dynamics without correction. (There is, however, one very conspicuous example of a failure of Newtonian physics: a very small error in predicting the advance of the perihelion of Mercury—40″ per century!—for which we need to invoke relativity theory.) Hence for engineering and all physics except a portion of atomic and subatomic physics, it is still the Newtonian physics that explains occurrences in the external world.

While it is true that the Newtonian mechanics is still applicable in the range of phenomena for which it was intended, the student should not make the mistake of thinking that the framework in which the system originally was set is equally valid. Newton believed that there *was* a sense in which space and time were "absolute" physical entities. Any deep analysis of his writings shows how in his mind his discoveries depended on these "absolutes." To be sure, Newton was aware that clocks do not measure absolute time, but only local time, and that we deal in our experiments with local space rather than absolute space. He actually developed not merely a law of gravitational force and a system of rules for computing the answers to problems in mechanics, but constructed a complete system based on a world view, encompassing ideas of space, time, and order. Today, following the Michelson-Morley experiment and relativity, that world view can no longer be considered a valid basis for physical science. The Newtonian principles are considered to be only a special, though extremely important, case of a more general system.

Some scientists hold that one of the greatest validations of Newtonian physics has been the set of predictions concerning satellite motions; they have enabled us to launch into orbit a series of space vehicles and to predict what will happen to them out in space. This may be so, but to the historian the greatest achievement of Newtonian science must ever be the first full explanation of the universe on mechanical principles—one set of axioms and a law of universal gravitation that apply to all matter everywhere: on earth as in the heavens. Newton recognized that the one example in nature in which there is pure inertial motion going on and on and on, without frictional or other interference to bring it to a halt, is the orbital motion of moons and planets. And yet this is not a uniform or unchanging motion along a single straight line, but rather along a constantly changing straight line, because planetary motions are a compounding of inertial motion with a continuing falling away from it. To see that moons and planets exemplify pure inertial motion required the same genius necessary to realize that the planetary law could be generalized into a law of universal attraction for all matter and that the motion of the moon partakes of the motion of the falling apple.

Isaac Newton's system of mechanics came to symbolize the rational order of the world, functioning under the "rule of nature." Not only could Newtonian science account for present and past phenomena; the principles could be applied to the prediction of future events. In the *Principia* Newton proved that comets are like the planets, moving in great orbits that must (according to Newtonian rules) be conic sections. Some comets move in ellipses and these must return periodically from far out in space to the visible regions of our solar system, whereas others will visit our solar system and never return. Edmond Halley applied these Newtonian results to an analysis of cometary records of the past and found—among others—a comet with a period of some seventy-five and a half years. He made a bold Newtonian prediction that this comet would reappear in 1758. When it did so, right on schedule, though long after Halley and Newton were dead, men and women everywhere experienced a new feeling of awe for the powers of human reason abetted by mathematics. This new respect for science was expressed by such adjectives as "amazing," "phenomenal," or "extraordinary." This successful prediction of a future event symbolized the force of the new science: the perfection of the mathematical understanding of nature, realized in the ability to make reliable predictions of the future. Not surprisingly, men and women everywhere saw a promise that all of human knowledge and the regulation of human affairs would yield to a similar rational system of deduction and mathematical inference coupled with experiment and critical observation. The eighteenth century not only was the Enlightenment, but became "preeminently the age of faith in science." Newton became the symbol of successful science, the ideal for all thought—in philosophy, psychology, government, and the science of society.

Newton's genius enables us to see the full significance of both Galilean mechanics and Kepler's laws of planetary motion as manifested in the development of the inertial principles required for the Copernican-Keplerian universe. A great French mathematician, Joseph Louis Lagrange (1736–1813), best defined Newton's achievement. There is only one law of the universe, he said, and Newton discovered it. Newton did not develop modern dynamics all by himself but depended heavily on certain of his predecessors; the debt in no way lessens the magnitude of his achievement. It only emphasizes the importance of such men as Galileo and Kepler, and Descartes, Hooke, and Huygens, who were great enough to make significant contributions to the Newtonian enterprise. Above all, we may see in Newton's work the degree to which science is a collective and a cumulattive activity and we may find in it the measure of the influence of an individual genius on the future of a cooperative scientific effort. In Newton's achievement, we see how science advances by heroic exercises of the imagination, rather than by patient collecting and sorting of myriads of individual facts. Who, after studying Newton's magnificent contribution to thought, could deny that pure science exemplifies the creative accomplishment of the human spirit at its pinnacle?

SUPPLEMENT NOTE ON THE TWO FORMS OF NEWTON'S SECOND LAW

Newton's *Principia* contains two forms of the second law. Since Newton's day we usually consider only the case of a continuously acting force F acting on a body of mass m to produce an acceleration A, that is $F = mA$. But Newton gave primacy to another case, that of an instantaneous force—an impact or blow—as when a tennis racquet strikes a ball or one billiard ball strikes another. In such cases, the force does not produce a continuous acceleration, but rather an instantaneous change in the body's quantity of motion (or momentum). This is the "change in motion" which is said to be proportional to "the motive force impressed" in Newton's statement of Law II on page 152. Newton conceived that $F = mA$ is a limiting case of the impact law, the situation when the time between successive impacts decreases indefinitely, so that the force ultimately achieves the limiting condition of acting continously. The law $F = mA$ was thus considered by Newton as derived from the impact law, as stated on page 152.

SUPPLEMENT
1

Galileo and the Telescope*

Galileo certainly did not invent the telescope and never claimed to have done so. Nor was he the first observer to point such an instrument toward the heavens. A newsletter of October 1608, about a year before Galileo made his first instrument, carried the news that the spyglass not only could make distant terrestrial objects seem nearer, but enabled one to see "even the stars which ordinarily are invisible to our eyes." There is very good evidence that Thomas Harriot had been observing the moon before Galileo began his telescopic observations; Simon Marius's claims (e.g., that he had discovered the satellites of Jupiter) are less well founded.

Galileo's report (see pp. 56–57) is taken from his *Sidereus nuncius* (1610). He wrote other versions of his first encounter with the telescope, which differ somewhat in detail, for instance, with

*This supplement is based on a report on this topic, by Albert Van Helden, at an international congress on Galileo, held in Pisa, Padua, Venice, and Florence in April 1983, published in the proceedings of this congress, edited by Paolo Galluzzi: *Novità celesti e crisi del sapere* (Suppl. to *Annali dell'Istituto e Museo di Storia della Scienza,* Florence, 1983). See also the monograph by Van Helden in the Guide to Further Reading on p. 243 below.

In *The Sidereal Messenger,* Galileo states that he had only heard of the new device, but had not actually seen one, when he applied his knowledge of the theory of refraction to produce a spyglass. But, by this time, the new instruments were not uncommon in Italy, and one had already arrived in Padua and was being discussed. Perhaps he was in Venice when the spyglass was being shown in Padua. In *The Assayer (Il Saggiatore)* of 1623, he recounted the role he played in the creation of the astronomical telescope and discussed in full the stages that led him to reinvent this instrument. Here, however, we are less concerned with the invention of the telescope than with the use Galileo made of it.

respect to his knowledge of the construction of the instrument (that is, the combination of two lenses, one plus and the other minus). What is of most significance is not that Galileo knew (or did not know) of the kinds of lenses needed to make such a telescope or spyglass, but that he very quickly made telescopes far superior in magnifying power and in quality to any others, telescopes good enough to serve the purpose of astronomical discovery. In this sense, Galileo transformed the crude spyglass into a refined astronomical telescope.

Galileo's contemporaries who made or sold spyglasses used common spectacle-makers' lenses, achieving very low magnification (some three or four times). Even Thomas Harriot, who was apparently in possession of spyglasses much before Galileo, was only able to get up to a six-power instrument by August 1609, at which time Galileo (who had just heard of the instrument in that month or in the previous month of July) had already made an eight-power or maybe a nine-power instrument. By the end of the year, he had reached twenty-power and had introduced an aperture ring to improve the image.

Not only did Galileo grind his own lenses of greater power than those used by spectacle-makers, but his lenses were of higher quality and his instruments had the advantage of the novel feature of an aperture ring. Albert Van Helden, the foremost scholar of this subject, concludes: "Even though Harriot preceded him in lunar observations [with the new instrument], Galileo was probably the first fully to grasp the meaning of the lunar features, the Moon's earth-like nature." By March 1610, Galileo had discovered stars never seen before, the difference in appearance between planets (showing a disk through the telescope) and fixed stars (appearing as scintillating points of light), the stars composing the Milky Way, and the satellites of Jupiter. These discoveries were published in the *Sidereus nuncius* in spring 1610. By July he had discovered protuberances on Saturn and later in the year the phases and correlated variations in size of Venus.

Galileo, in fact, found almost all there was to be discovered with this type of telescope—being first to do so in part because he was far in the lead in having an adequate instrument. But by 1611 others had obtained telescopes that enabled them to distinguish celestial phenomena, even though (as Van Helden points out) their telescopes were probably not as good as Galileo's. So it is that there are rival claimants to the discovery of sunspots in 1611. Van Helden comments that this was "the last major discovery of this initial phase of telescopic astronomy." Further discoveries of significance would require higher magnification and a resolution beyond the ability of the lenses of this first period.

Until the 1630s, Galileo was still making and distributing telescopes. But the next decades witnessed the emergence of new instruments, no longer composed of a single negative lens as eyepiece and a single positive lens as objective. In the 1630s, other astronomers produced maps of the moon and studies of sunspots, observed the transit of Mercury in 1631 and of Venus in 1639, and found markings on the surface of Jupiter. Galileo did not participate in these further developments.

The "second wave of discoveries" with new telescopes can be dated as beginning in 1655, with Huygens's discovery of Titan, a satellite of Saturn. Huygens later was able to resolve Galileo's puzzling observations of the protuberances of Saturn. He found them to be a planar ring encircling the planet.

Galileo's major contribution to the telescope has been summed up as follows: he changed "a feeble spyglass into a powerful research instrument." He was first in being able "to grind long-focus objective lenses" (that were of good quality) and he was the first to equip his instruments with aperture rings. In short, he was the first scientist to attain "astronomically significant magnifications at acceptable qualities." Van Helden concludes that Galileo "discovered single-handedly all the important things that could be discovered with this generation of instruments, except for sunspots, which were discovered independently by several other observers."

SUPPLEMENT
2

What Galileo "Saw" in the Heavens*

An analysis of Galileo's experience on looking at celestial objects through the telescope in 1609 and succeeding years shows how his commitment to the Copernican doctrines conditioned and even, to some degree, directed the interpretation of what he actually observed. Writers on the history of science often convey the impression that in 1609 Galileo discovered or "saw" mountains on the moon and satellites of Jupiter. A careful reading of Galileo's manuscript records or the published account of his discoveries that he presented in his *Starry Messenger* of 1610 shows, however, that when Galileo examined the moon through the telescope, what he actually saw was a larger number of spots than he had expected. Some of the spots were darker and very much bigger than others; they were called by Galileo "the 'large' or 'ancient' spots," since these were the ones seen and reported by naked-eye observers during many centuries. They were to be distinguished from certain smaller and very numerous spots that had never been observed until the invention of the telescope—or as Galileo said, "had never been seen by anyone before me." These new spots were the raw data of sense experience. Or, to put it another way, what Galileo actually *saw* through the telescope was a collection of spots of two sorts. It took some time until, as Galileo tells us, he transformed these sense data or visual

*This supplement is based on my monograph on "The Influence of Theoretical Perspective on the Interpretation of Sense Data" in *Annali dell'Istituto e Museo di Storia della Scienza di Firenze,* anno V (1980), fascicolo 1. The translations from Galileo's *Sidereal Messenger* are from Stillman Drake's *Discoveries and Opinions of Galileo* (Garden City, N.Y.: Doubleday & Co., 1957).

images into a new concept: a lunar surface with mountains and valleys, the source and cause of what he had seen through the telescope. On this score there can be no doubt whatever, as Galileo himself made clear in his published account. Let him speak for himself:

> From observations of these spots repeated many times I have been led to the opinion and conviction that the surface of the moon is not smooth, uniform, and precisely spherical as a great number of philosophers believe it (and the other heavenly bodies) to be, but is uneven, rough, and full of cavities and prominences, being not unlike the face of the earth, relieved by chains of mountains and deep valleys.

Then Galileo describes the actual observations he had made "by which I was enabled to draw this conclusion." We may note that many of them suggested to Galileo's mind an analogy with terrestrial phenomena. For example, certain "small blackish spots" had "their blackened parts directed toward the sun" while on the side opposite the sun they appeared to be "crowned with bright contours, like shining summits." We see a similar phenomenon on earth at sunrise, Galileo remarks, "when we behold the valleys not yet flooded with light though the mountains surrounding them are already ablaze with glowing splendor on the side opposite the sun." Another "astonishing" observation was a series of "bright points" in the dark region of the moon well beyond the terminator. He found that these would gradually get larger and eventually join the "rest of the lighted part [of the moon] which has now increased in size." These, he concluded, must be luminous mountain peaks rising so high from the surface of the moon that they are illuminated by the sun's light, even though their bases are in the region of shadow or in darkness. Again Galileo reminds his reader of a terrestrial analogy, since "on the earth, before the rising of the sun, are not the highest peaks of the mountains illuminated by the sun's rays while the plains remain in shadow?"

The intellectual transformation of these lunar observations into conclusions that agree with what Galileo calls "the old Pythagorean opinion that the moon is like another earth" was propelled by Galileo's commitment to the Copernican system. There must have been an enormous unconscious pressure to vindicate the Copernican position that the earth is merely another planet, that it is not fundamentally different from the other planets and the moon. If the earth is not a unique body, it is not specially conditioned to be motionless and at the center of the universe. Galileo's commitment to Copernicanism thus impelled him to transform the data of observation into an argument that the moon resembles the earth.

A somewhat similar process of transformation of the sense data of experience occurred in relation to what Galileo called "the matter which in my opinion deserves to be considered the most important of all—the disclosure of four Planets never seen from the creation of the world up to our time." In this declaration, Galileo is using the word "planet" in the original Greek sense of any wandering body in the heavens and is referring to his discovery of satellites of Jupiter, or secondary planets accompanying the primary planet Jupiter. What he actually "saw" was not a set of moons or satellites. He actually observed, on 7 January 1610, "beside the planet . . . three starlets, small indeed, but very bright." These points of light, looking like stars despite their proximity to Jupiter, were the actual sense data. Galileo at first made only the simple and obvious transformation of the sight of these points of light and concluded that he had seen stars. As he says, "I believed them to be among the host of fixed stars." The only special aspects that aroused his curiosity, he goes on, were their "appearing to lie in an exact straight line parallel to the ecliptic, and . . . their being more splendid than others of their size." So far was he from conceiving that these might be satellites of Jupiter that he tells us that he "paid no attention to the distances between them and Jupiter, for at the outset I thought them to be fixed stars, as I have said." His second observation occurred on the next night and showed "three starlets . . . all to the west of Jupiter, close together, and at equal intervals from one another." Even then, Galileo did not begin to guess that these were satellites. Rather, he tells us,

> I began to concern myself with the question of how Jupiter could be east of all these stars when on the previous day it had been west of two

of them. I commenced to wonder whether Jupiter was not moving eastward at that time, contrary to the computations of the astronomers, and had got in front of them by that motion. Hence it was with great interest that I awaited the next night.

After further observations, he eventually "decided beyond all question that there existed in the heavens three stars wandering about Jupiter as do Venus and Mercury about the sun." Before long he had found that "four wanderers complete their revolutions about Jupiter." It is not without interest that Galileo draws an analogy between the satellites or lesser lights moving around the greater light of Jupiter and the motion of Venus and Mercury about the brighter light of the sun. This analogy would indicate that Galileo's Copernicanism was directly related (according to his own testimony in so many words) to his transformation of the idea that there are *stars* moving *along with* Jupiter into the idea that there are *satellites* moving *around* Jupiter.*

The example of the satellites of Jupiter differs in one essential from the earlier experience with spots on the moon. Galileo's Copernicanism and anti-Aristotelianism obviously preconditioned his mind to the possibility that the moon would be earthlike. But there was nothing in his anti-Aristotelian bias or his pro-Copernican commitment to prepare him for the existence of a model of the Copernican system in miniature in the form of a satellite system around Jupiter. Looking backward, after the events, the following reasoning must seem likely: If the earth is not unique, then it should follow that the earth is not the only planet with a satellite. This line of thought might possibly have been a part of Galileo's ultimate conception that there are satellites of Jupiter. But in fact Galileo does not mention the analogy

*On Galileo's actual observations and new "light on Galileo's actual process of reaching the conclusion that he was seeing bodies which literally circulated around Jupiter," see Stillman Drake, *Galileo at Work: His Scientific Biography* (Chicago and London: The University of Chicago Press, 1978), 146–53, esp. 148–49.

Drake has also shown what a labor of heroic proportions it was for Galileo actually to determine the periods and orbital radii (or maximum elongations) of Jupiter's satellites. There must have been a strong commitment to the concept of satellites for Galileo to undertake such an "atlantic labour." See Drake, "Galileo and Satellite Prediction," *Journal for the History of Astronomy* 10 (1979), 75–95.

with the earth's having a moon. In any event there is an astonishingly great difference between a planet's having a single moon and the existence of a whole satellite system of four new "planets" encircling Jupiter. Even so firm a Copernican as Kepler was shattered by the news that Galileo had discovered four new planets or wandering stars, since he did not quite know how he could fit them into his scheme in which the separation between six planets was related to the existence of five, and only five, regular geometric solids.

Of course, there was an additional point to the new discovery, once it had been made, and that was that it answered the questions of the anti-Copernicans who argued that the earth could not move in its orbit (and remember that it does so at the enormous speed of about twenty miles per second) without losing its moon. Everyone admitted that Jupiter must move; well, if Jupiter could move in orbit and not lose four moons, surely there could be no objection to the earth's moving and not losing its single moon!

Before long, Galileo (and others) had made another remarkable discovery, namely, that the sun has spots. These spots are the given, the data of sense observation. What is significant is how they were transformed or interpreted by the mind of Galileo. It is well known that Galileo showed these to be actual spots on the surface of the sun and thus interpreted their motion as an indication that the sun rotates on its axis. Others, who were of a different scientific and philosophical point of view, tried to give another interpretation, holding that these were shadows cast on the sun, possibly by "stars," either "fixed" or "wandering" (i.e., "planets"), that "revolve about it in the manner of Mercury or Venus." The two interpretations show how different points of view interact in the mind of a scientist with what is observed. An Aristotelian must believe that the sun itself is pure and unspotted, whereas an anti-Aristotelian like Galileo did not care whether the sun is spotted or unspotted, whether it is immutable or undergoes changes from day to day. The sunspots are of interest in the present context in a historical sense, because it turns out that there had been in the Middle Ages a certain number of observations of sunspots, but these had tended to be interpreted as instances of the passage of a planet (Mercury or

Venus) across the face of the sun, since the prevailing philosophy would not permit these observations to be transformed into the interpretative statement that the sun has spots on it.*

The doctrine of transformation tends to pinpoint the actual occasion on which the scientist's background, philosophical orientation, or scientific outlook interacts with sense data in order to provide the kind of base on which science advances. The next phase of investigation would be to identify, classify, and interpret, in a number of examples, those parts of scientists' background that are operative in discoveries. A first assignment would be to try to distinguish between the effect of the general background in philosophy and science and the effect of the particular personality of the scientist. It would be important to try to find the degree to which intellectual transformations are either related to the background or are independent of the particular scientist. Only the barest beginnings have been made in this general area of the psychological background to discovery. In particular, this was the subject of a very perceptive set of observations by N. R. Hanson, and it has been explored by Leonard K. Nash. Gestalt psychology may have much to contribute here. And there is no doubt that studies by experimental psychologists such as R.L. Gregory and art historians such as E. H. Gombrich will eventually do much to illuminate this topic.**

*On the sunspot debate, see Stillman Drake's translation of Galileo's *History and Demonstrations Concerning Sunspots and Their Phenomena,* 59–144, esp. 91–92, 95–99. Bernard R. Goldstein has written "Some Medieval Reports of Venus and Mercury Transits" in *Centaurus* 14 (1969), 49–59.

**Norwood Russell Hanson, *Patterns of Discovery: An Inquiry into the Conceptual Foundations of Science* (Cambridge: at the University Press, 1958); Leonard K. Nash, *The Nature of the Natural Sciences* (Boston: Little, Brown and Company, 1963). On the question of Gestalt in relation to scientific discovery, see (in addition to the works of Hanson and Nash) Thomas S. Kuhn, *The Structure of Scientific Revolutions,* 2d ed. (Chicago: The University of Chicago Press, 1970), 64, 85, 111, 122, 150; and Kuhn, *The Essential Tension* (Chicago: University of Chicago Press, 1977), xiii. Also see R.L. Gregory, *The Intelligent Eye* (London: Weidenfeld and Nicolson; New York: McGraw-Hill Book Co., 1970); Gregory, *Eye and Brain: The Psychology of Seeing* (New York: McGraw-Hill Book Co. [World University Library], 1966); R.L. Gregory and E.H. Gombrich, eds., *Illusion in Nature and in Art* (New York: Charles Scribner's Sons, 1973); and E.H. Gombrich, *Art and Illusion* (New York: Pantheon Books, 1960).

SUPPLEMENT
3

Galileo's Experiments on Free Fall

In some unpublished writings from his Pisa days, Galileo describes experiments of dropping unequal weights from a tower. He does not identify which tower he used, but I assume that it must have been the famous Leaning Tower. When I repeated this experiment, in the company of a group of scholars assembled in Florence and Pisa for an International Congress of the History of Science, I discovered that because of the architectural features of the tower, it was necessary to lean way out with arms held horizontally, a weight in each hand. Clearly, the result of the experiment—whether the weights arrive at the ground simultaneously or at differing times—depends on the degree of simultaneity of release of the weights. Galileo's notes indicate that sometimes a heavy ball would start out more slowly than a light ball, but would then overtake it during the descent. This seemed a puzzling result—and all the more so in that it occurs in his own unpublished manuscripts, which we may assume contain true and unbiased records of what was actually observed. In other cases, Galileo reported that two unequal weights fall almost simultaneously, or that there is only the small difference that he mentions in the *Two New Sciences.*

If Galileo was a careful experimenter, then what of the results concerning the light ball moving ahead of the heavy one? It is certainly to Galileo's credit that he recorded this phenomenon (and said he had observed it "many times"); he even tried to explain this strange occurrence, which did not quite fit his theories. Galileo's statements are unambiguous. He wrote that "if an observation is made, the lighter body will, at the beginning of the motion, move ahead of the heavier and will be swifter." Again, if two spheres of equal size, one twice the weight of the other, are "dropped from a tower," it will be found that at the beginning of the motion "the lighter one will move ahead of the heavier, and for some distance will move more swiftly." Galileo even tried to account for this phenomenon in a chapter of his unpublished tract on motion entitled "In which the cause is given why, at the beginning of their natural motion, bodies that are less heavy move more swiftly than heavier ones." Not only does Galileo assert that he has observed this phenomenon, but he cites a similar observation made by Girolamo Borro in a book of 1575; Borro was a professor at Pisa, still teaching there during Galileo's student days. Borro dropped pieces of wood and of lead of equal weight but of unequal size and found that the "lead descended more slowly." He writes that trial was made "not only once but many times," and "with the same result."

We are grateful to Thomas B. Settle* for a solution to this puzzle. He reports that when an experimenter holds two unequal weights, palms downward, with outstretched arms, it is not possible to release the two weights simultaneously. Even though the experimenter fully believes that the two have been released at the same instant, photographic evidence shows incontrovertibly that the hand holding the heavy weight invariably opens a short time after the one holding the light weight. This is apparently an effect of differential muscular fatigue depending on the magnitude of the weight. The discovery that in this case, as in others, the results of experiment accord with Galileo's reports gives us confidence in Galileo as a gifted experimenter who recorded and reported exactly what he observed.

Furthermore, this episode provides additional evidence that Galileo was making experiments with freely falling objects very early in his career and that experiments were of real significance in his exploration of the science of motion.

*"Galileo and Early Experimentation," in Rutherford Aris, H. Ted Davis, and Roger H. Stuewer, eds., *Springs of Scientific Creativity* (Minneapolis: University of Minnesota Press, 1983), 3–20.

SUPPLEMENT

4

Galileo's Experimental Foundation of the Science of Motion

Until fairly recently, our knowledge about Galileo's studies of motion was based on his tracts and books (those published by him during his lifetime and those later edited and published after his death), manuscript notes, and correspondence. These were assembled into a magnificent twenty-volume edition, under the general editorship of Antonio Favaro (1890–1909; reprinted 1929–1939, and again 1964–1966). From these materials there emerged a record of development of ideas leading from Galileo's early thinking in a late medieval mode of impetus physics to his discovery of the laws of free fall (that there is a constant acceleration that produces a velocity increasing as the time and a distance increasing as the square of the time) and his brilliant application of the principles of resolution and composition of vector velocities in order to analyze the trajectories of projectiles.

In the decades following World War II, many scholars—following the lead of Alexandre Koyré—had concluded that in the stages of discovery and development of the principles of motion, the role of true experiment was minimal. Galileo was seen as a thinker and analyst, not one who put direct questions to the test of experience. It was even doubted that Galileo had ever performed the inclined-plane experiment described in the *Two New Sciences* as a confirmation of the conclusions arrived at by mathematical analysis. Most scholars agreed that the reported exactness of observations within "a tenth of a pulse-beat" far exceeded the capacity of this apparatus; here was apparent evidence that

Galileo had probably never done this experiment. The best that could be said for Galileo was that he had boastfully exaggerated the results. This point of view seemed all the more justified to the degree that Galileo gave no numerical data. Doubts concerning the inclined plane were not voiced for the first time in the twentieth century. In Galileo's own time, Father Marin Mersenne wrote in *Harmonie universelle,* 1 (Paris, 1636), 112: "I doubt that Galileo performed experiments on the inclined plane, because he never speaks of them and because the proportion given often contradicts the experimental evidence."

Today our view of the matter has undergone a radical change. In 1961, Thomas B. Settle devised and performed an experiment that closely replicated the one described by Galileo in the *Two New Sciences.* In his report ("An Experiment in the History of Science," *Science* 133 [1961], 19–23), Settle showed that the results were, just as Galileo said, easily accurate to within a tenth of a pulse-beat. Others confirmed Settle's results. Another experimenter, James MacLachlan (*Isis* 64 [1973], 374–79), then repeated an effect described by Galileo, which had been the subject of particular derision and had been used to underline the fact that Galileo's experiments were only "thought-experiments" and obviously could not possibly give the results described by Galileo. But MacLachlan found that this experiment, unbelievable at first encounter, accorded exactly with Galileo's description. We have seen (in Supplement 3) that in the early 1590s, while still at Pisa, Galileo was making experiments with falling bodies and that there is a reasonable explanation for the bizarre result he recorded that a light body starts out ahead of a heavy body when both are released "simultaneously."

The growing awareness of the experiments actually made by Galileo has led to a renewed interest in trying to ascertain, as precisely as possible, the path that Galileo followed in his discovery of the laws of motion. Were his steps directed primarily by intellectual analysis as his published works would have us believe? Or were his ideas developed in the course of performing experiments? In the early 1970s, Stillman Drake made a new study of Galileo's manuscripts. Among them he found pages that had been omitted by Favaro in his edition of Galileo because "they contained only calculations or diagrams without attendant propositions or explanations."*

Drake concludes his analysis of these data and diagrams by asserting that they include "at least one group of notes which cannot satisfactorily be accounted for except as representing a series of experiments designed to test a fundamental assumption, which led to a new, important discovery." The assumption, according to Drake, was that of linear inertia; the discovery was that a slowly moving projectile (a ball rolling down an inclined plane, striking a deflector and shooting out into space) has a curved path that looks like a parabola. Drake confirmed a hunch of Favaro that Galileo had discovered the parabolic trajectory of projectiles as early as 1609, and that furthermore at this time he knew and wrote out the proofs of the propositions concerning parabolic motions, essentially as they appear in the Fourth Day of the *Two New Sciences.* Drake has also analyzed some other manuscript sheets that contain data that accord with a mode of discovery of the law of free fall in the course of making experiments.

The new image of Galileo that emerges from Drake's studies is that of a scientist in the modern mode, exploring the subject of motion through experiments (much in the manner that physicists have been doing for the past two centuries) and committed to a philosophy of science similar to that adopted by many twentieth-century physicists. Drake minimizes Galileo's alleged dependence on medieval precursors, pointing to the absence of the concept of mean speed from Galileo's early writings and showing how Galileo used a new Eudoxian approach to the theory of proportions.

*Stillman Drake's analyses have been published in a number of articles, among them "Galileo's Experimental Confirmation of Horizontal Inertia," *Isis* 64 (1973), 291–305; "Galileo's Discovery of the Law of Free Fall," *Scientific American* 228, no. 5 (May 1973), 84–92; "Mathematics and Discovery in Galileo's Physics," *Historia Mathematica,* 1 (1974), 129–50; and, with James MacLachlan, "Galileo's Discovery of the Parabolic Trajectory," *Scientific American* 232, no. 3 (March 1975), 102–10. For a summary, see Drake's *Galileo at Work.*

Not all historians agree fully with all of Drake's analyses and conclusions.* One problem is that Drake seems overly committed to the image of Galileo as a modern (e.g., nineteenth-century or later) physicist, a man who broke with tradition, whereas many historians have tended to see Galileo as innovative but nevertheless having strong links with medieval and Renaissance thinkers.** Furthermore, Drake has not restrained his opinions. For instance, he states openly that "to find manuscript evidence that Galileo was at home in the physics laboratory hardly surprises me." And he openly attacks the received opinion "of our more sophisticated colleagues" who, he says, have advanced "philosophical interpretations" whose only merit is that they "fit with preconceived views of orderly long-term scientific development." Drake dislikes the view, which he considers pejorative, that "Galileo was an armchair speculator." He would have us believe rather that to a "remarkable extent" Galileo "made use of experimental methods in science that we now take for granted but that were not standard procedure in the 17th century."

*Two of the major scholars who have not agreed with Drake's conclusions are Winifred L. Wisan and R.H. Naylor. Winifred Wisan's most significant work is "The New Science of Motion: A Study of Galileo's *De motu locali,*" *Archive for History of Exact Sciences* 13 (1974), 103–306; see also "Mathematics and Experiment in Galileo's Science of Motion," *Annali dell'Istituto e Museo di Storia della Scienza di Firenze* 2 (1977), 149–60, and "Galileo and the Process of Scientific Creation," *Isis* 75 (1984), 269–86. R.H. Naylor's ideas are presented in his "Galileo and the Problem of Free Fall," *British Journal for the History of Science* 7 (1974), 105–34; "The Role of Experiment in Galileo's Early Work on the Law of Fall," *Annals of Science* 37 (1980), 363–78; and "Galileo's Theory of Projectile Motion," *Isis* 71 (1980), 550–70.

**Drake not only asserts that experiments played an essential role in Galileo's discovery of the laws of motion, and that his particular interpretation of Galileo's diagrams and data reveals Galileo's actual path of reasoning and analysis, but also denies the importance of the late medieval studies of motion for Galileo. The extent and significance of Galileo's knowledge of (and use of) ideas of fourteenth-, fifteenth-, and sixteenth-century concepts, principles, and methods are currently the subject of intensive historical study by, among others, William Wallace, Alistair Crombie, and Antonio Carrugo. An exploration of the medieval ideas on motion that tends to minimize the significance of this development for seventeenth-century physics has been written by John E. Murdoch and Edith D. Sylla (see Supplement 7).

Hence his results imply that Galileo's mode of presenting his principles and laws of motion differs radically from Galileo's path to discovery.

But even if scholars do not accept the details of Drake's individual analyses and his conclusions, there can be very little doubt that Drake's researches have shown that early in his scientific life Galileo was performing experiments on motion, and that such experiments were very closely related in some manner to his great discoveries. Drake has given us a version of the steps that led to Galileo's discovery of the laws of motion that reasonably fits Galileo's diagrams and numerical data. Yet it must remain an open question whether a somewhat different version also possibly explains these bare jottings. In the absence of explanatory notes or comments by Galileo himself, any reconstruction must be somewhat tentative and hypothetical. In order to make his reconstructions and to give Galileo's numbers, diagrams, and occasional notes a physical significance, Drake has had to make a number of assumptions, to guess at possible intermediate stages of thought. What emerges is a consistent picture, but one that has not been universally accepted.

In the context of our study of the birth of a new physics, however, we may minimally conclude that Drake has proved in Galileo's case that there was an enormous difference between what Alfred North Whitehead once called the "logic of discovery" and the "logic of the discovered." Drake's analysis of Galileo's manuscript pages indicates that experimental investigations must have played an essential role in the "logic of discovery," the way in which Galileo may have obtained his results. Since Galileo's published presentation does not include such an experimental base, it must therefore constitute a "logic of the discovered," a recasting of the subject by Galileo so that the order and mode of presentation of his new science of motion would follow some preferred logical sequence. Be that as it may, it remains a fact of history that during the four centuries from Galileo's day to our own, the development of science and of scientific thinking influenced by Galileo has had to depend on the presentation he bequeathed to us in his *Two New Sciences.* *

There are, however, a set of factors that tend to favor Drake's arguments in addition to the fact that his reconstruction does fit the numbers on Galileo's manuscript pages and the diagrams. These additional factors are largely negative; that is, they provide evidence that Galileo's path to discovery—even if it should turn out to differ in a major way from Drake's proposal—could not have been the tidy analysis he presented in the *Two New Sciences.* First of all, Galileo's early writings on motion do not use the concept of a continuous acceleration, which appears so prominently in the *Two New Sciences* and which he presumably could have learned from the late medieval writings on motion. In his earliest treatise on motion (composed at Pisa around 1590), he explored speeds along inclined planes, concluding incorrectly that along planes of different lengths but equal heights, the speeds ought to be proportional to the lengths of the planes. At that time, he evidently considered acceleration to be only a minor effect at the beginning of the motion, and not to be operative continuously. Nor does the true concept of acceleration appear in a treatise on mechanics composed in 1592, shortly after he moved to Padua. By 1602, he had found that the time would always be the same for a body "falling" freely along any chord of a vertical circle that ended in the circle's lowest point. But in his discussion of this result, he again does not discuss acceleration. It was not until 1603–4 that Galileo began to concentrate on the concept of acceleration in his search for a rule that would account for free fall in terms of distances, speeds, and times. Now the time in Galileo's career when he was most aware of the late

*Many scientists and historians of the late nineteenth and early twentieth centuries uncritically assumed that since Galileo was the "father" of modern physics (if not of modern science), he was equally the inventor and initiator of the experimental method. It followed that he must have made all of his discoveries by experiment. So prevalent was this view that the translators of Galileo's *Two New Sciences,* Henry Crew and Alfonso de Salvio, added the words "by experiment" to Galileo's text, so that his introduction to the subject of motion would not merely refer to the principles Galileo himself said he "found" *(comperio,* "I find"*),* but would make him say that these were new principles which "I have discovered by experiment."

medieval analysis of motion would almost certainly have been at the start of his investigations, when he was a young teacher at the university, not too long after he had ended his own period as a student. And yet this is just the time when the concept of acceleration and its consequences seem conspicuously absent from his thinking or of little importance. Hence he appears to have been following an independent path of exploration and discovery and not merely applying earlier results.

Furthermore, a key concept of the analysts of motion of the fourteenth, fifteenth, and sixteenth centuries was "mean speed," which figured prominently in the mean-speed theorem or the law of the mean. Again, it seems that this concept is not to be found in Galileo's early writings. In *Two New Sciences,* which is his last work, we do find a presentation much like that of the mean-speed theorem, but it is developed by Galileo in a way that is somewhat different, as a close analysis reveals. Even if it may be argued that Galileo came upon the medieval theory of motion later in life, that is, after his writings of 1590–1602, it would still be a mystery why the concept of the mean speed does not appear in a prominent position in his most mature writings. Thus there is some evidence that the stages of development of Galileo's thoughts on motion did not simply follow the line of the medieval thinkers.

It has been mentioned that one of the criticisms leveled at Drake is that he has had to introduce certain assumptions or hypotheses for which there is no direct evidence whatever, and that in doing so he may have been overly motivated by a desire to project an image of Galileo as an essentially modern kind of physicist of a special kind. Drake makes no secret of this image of Galileo and he is quite open about his suppositions. I myself tend to favor many of Drake's analyses, though I am troubled by some of the ancillary hypotheses needed to make his conclusions fit the recorded data. But I strongly dissent from Drake's polemical stance, expressed by him in these words:

> It is quite true that if Galileo had started out with a correct definition of uniform acceleration, as he did in his final published book, he would have been led ineluctably to his conclusions; and it is also true that such a definition had been given in the Middle Ages. All he would have

> needed to do would be to have applied this definition to the case of free fall, and of course to have added the postulate concerning speeds at the ends of inclined planes, which seems really to have been rather trivial and easy. And it is thus that Galileo's work is presented in textbooks, as a rather humdrum extension of medieval analyses of motion.

If I take exception to such a statement, I do so as a critic who has been a friend and admirer of the author for decades. Note the second sentence, which begins, "All he would have needed to do . . ." During two whole centuries (fourteenth, fifteenth), not one of the writers on this topic ever "applied this definition to the case of free fall," and in the next (sixteenth) century only one writer did so, but in a trivial way that had no effect or influence whatever. Hence, the evidence proves that to have made such an application must have been a heroic and tremendous step that had never been taken by any of the great philosophers (or natural philosophers), theologians, or mathematicians concerned with this topic. It is simply unhistorical to refer to this gigantic step, which in fact required a wholly new and revolutionary attitude concerning mathematics and nature, and the correlative step of the postulate concerning speeds at the ends of inclined planes, as seeming "really to have been rather trivial and easy." And, even if Galileo's use of medieval concepts and laws of uniform and uniformly accelerated motion comes only in his final presentation and does not account wholly for his discoveries, it is surely a gross distortion to refer to "a rather humdrum extension of medieval analyses of motion."

To sum up: Drake has called attention to some fundamental problems in accepting the analysis of motion given by Galileo in the *Two New Sciences* as if it were a true and complete account of his stages of discovery. Additionally, Drake has incontrovertibly shown that early in his career Galileo was experimenting with motion in the course of discovering his famous laws. Drake has also given an alternative reconstruction of Galileo's thought that accords with the data, if we are willing to grant certain not unreasonable suppositions. But there are some legitimate grounds for not giving full assent to every part of this reconstruction and for wondering to what degree what we have now been given is indeed the only scenario possible. Perhaps an alternative set of suppositions may combine the diagrams and data with at least some aspects of the presentation Galileo himself gave us in the *Two New Sciences.* But there are no grounds for doubt that experiment played a significant role in Galileo's studies of the principles of motion and the discovery of the laws of motion, in a way that had no basis in evidence until Drake's researches among the manuscripts in 1972. It must be said, in conclusion, that Drake's reconstruction *does* fit the diagrams and the data. It is therefore reasonable to conclude that Drake is essentially correct, even if with the zeal of the discoverer he may somewhat overstress his own image of Galileo as a modern experimental physicist and downplay both the role of intellection and the debt of Galileo to any of the late medieval concepts and rules concerning motion. In this case we have to accept the curious situation in which Galileo not only presented his results in the *Two New Sciences* in a wholly different way from the way in which he had discovered them, but effectively hid any trace of the steps that had led him to those discoveries. In this account he introduced an experiment (or the evidence of experiment) not in relation to discovery, as we have seen, but only as a test or check that the $D \propto T^2$ relation occurs in nature. And this appears in a section of the *Two New Sciences* in which Galileo introduces the subject of motion with a reference to the new things "I find" *(comperio).* It has taken some three and a half centuries for scholars to discover and to study Galileo's odd pages of jottings and calculations and to begin to penetrate beyond the logical facade of the *Two New Sciences,* so as to find the first steps of discovery and innovation.

SUPPLEMENT
5

Did Galileo Ever Believe that in Uniformly Accelerated Motion the Speed Is Proportional to the Distance?

In a study of "Galileo's Work on Free Fall in 1604" (in *Physis* 16 [1974], 309–22), Stillman Drake has discussed the letter Galileo wrote to Paolo Sarpi, in which he asserted that if it can be assumed that speeds in free fall increase as the distance, he could prove that the distances are as the square of the time. Drake's analysis is based on manuscript pages of Galileo's jottings. Drake concludes that Galileo was measuring a *velocità* by impact effects, as in the action of a pile driver, a quantity that would be akin to our V^2 rather than V. If, then, we were to translate Galileo's conditional statement to Sarpi into the language of algebraic proportions, Galileo would be made to say that the condition

$$V^2 \propto D$$

leads to the relation

$$D \propto T^2.$$

It is readily seen that this is merely another aspect of the fundamental relation

$$V \propto T$$

or

$$V^2 \propto T^2.$$

In the *Two New Sciences,* however, Galileo rather explicitly admits that he had once believed that

$$V \propto S$$

and only later was converted to the correct principle, that

$$V \propto T.$$

Sagredo (in the dialogue of the Third Day) asks whether "uniformly accelerated motion" is not "that in which the speed goes [on] increasing according to the increase of space traversed?" The reply, by Salviati (who generally speaks for Galileo), is that he finds it "very comforting to have had such a companion in error" and that "our Author himself . . . labored for some time under the same fallacy." Simplicio (the Aristotelian member of the group of discussants) adds his voice: he too believes "the speed [to be] increasing in the ratio of the space."

SUPPLEMENT

6

The Hypothetico-Deductive Method

In his experimental test, Galileo displays the essence of what has been called both the mathematico-experimental and the hypothetico-deductive method. He wishes to test the relation $V \propto T$, but there is no way for him to make a direct correlation of speeds and times determined by experiment. He knows, however, that if $V \propto T$, then $D \propto T^2$; that is, $D \propto T^2$ can be deduced from the hypothesis $V \propto T$. Also, he knows that he can make an experimental test of $D \propto T^2$. The confirmation of this relation gives him confidence that $V \propto T$, from which $D \propto T^2$ was derived, is valid.

In symbolic terms, what Galileo did was to deduce B from A; he next tested B, and then concluded that A holds. It should be noted, however, that this method does not include a guarantee of A. For instance, it might happen that B could also be deduced from A′. Additionally, it is assumed that the process of deducing B from A is correct. Traditionally, this means correctness of logical deduction. Galileo's method is to derive B from A by the aid of mathematics. Because B is derived from A by mathematics and then tested by experiments, the method can also be called mathematico-deductive. In the seventeenth century, the term "mathematical-experimental" was also used. This method is called "hypothetico-deductive" because we wish to test *hypothesis A* but cannot do so by direct experiment. Hence we *deduce B* from A and then test the *deduction B* by experiment. (Galileo's use of this method in relation to the hypothesis $V \propto T$ and the testable deduction $D \propto T^2$ may be found on pp. 88 ff. *supra.*)

SUPPLEMENT
7

Galileo and the Medieval Science of Motion

In attempting to relate the growth of Galileo's ideas on motion to the late scholastic analysis, care must be taken to distinguish any use Galileo might have made of the work of these predecessors in the course of discovery (see Supplement 4) and in his logical presentation of this subject in the *Two New Sciences.* Furthermore, it must be kept in mind that the medieval writers were dealing with abstractions and not the world of nature as revealed by our senses and known to us by experiment and observation. John Murdoch and Edith D. Sylla* have summed up the matter as follows:

> Even when the causes of a motion were present and were measured as the forces and resistances determining that motion, the concern was not with these forces and resistances as dwelling in some particular mover, mobile, or medium, but with the forces and resistances in abstraction from concrete agents and patients.

Hence, it is an error to regard the "new and distinctive fourteenth-century efforts as moving very directly toward early modern science." While "Galileo was aware of medieval works on motion," and may have put the "medieval mean-speed theorem and even its proof to his own use in his investigation of naturally accelerated motion," it must be remembered that "what was then being used was but one part, one fragment, of the medieval

*John E. Murdoch and Edith D. Sylla, "The Science of Motion," in David Lindberg, ed., *Science in the Middle Ages* (Chicago: The University of Chicago Press, 1978), 206–64.

'science of motion,' a part removed from its context and, in Galileo's hands, made to perform quite different duty."

In short, "the goal, indeed, the whole enterprise, of many a medieval scholar who treated motions was worlds away from that of Galileo and his confreres." Even the mean-speed theorem was "never (save once, almost by accident) put together with the motion of free-fall, as was the case with Galileo."

These strictures serve to remind us that there could have been no easy transition from these late scholastic writings to Galileo's new and revolutionary science of motion. There is, in fact, no better index to the truly revolutionary quality of Galileo's new science of motion than a contrast between these medieval abstractions divorced from any taint of nature and the Galilean science based squarely on observations and experiments and tested by its degree of conformity to nature as revealed by experiment.

SUPPLEMENT
8

Kepler, Descartes, and Gassendi on Inertia

In this presentation, I have omitted the contributions of Kepler, Descartes, and Gassendi. Kepler introduced the term *inertia* into the discourse on motion. But for Kepler inertia (from the Latin word for laziness or indifference) implied primarily that matter cannot by and of itself either begin to move or remain in motion if moving. Rather, because of its "inert-ness," matter needs a mover. Whenever the mover stops acting, the body must come to rest, and must do so wherever it may happen to be. By itself and without a mover a body would not continue to move to some "natural place." Thus Kepler's physics implied that there can be no natural places which bodies can seek, as Aristotle had taught. This radical conclusion was necessary for Kepler, since in the Copernican universe the earth is in constant motion and so there can be no fixed or natural place for terrestrial bodies.

Descartes had a much more radical idea. He put forth the concept that uniform straight-line motion is a "state," just as rest is. Because a body can maintain itself in any "state" without the action of an external force, Descartes was in essence making a dynamical equivalence of a state of rest and a state of motion, so long as the latter is uniform and rectilinear. Descartes first expounded this new principle in a work entitled *Le monde,* that is, *The World* or *The Universe.* But he did not publish this treatise, which was composed on a Copernican basis. When Descartes learned of the condemnation of Galileo by the Roman Inquisition, he decided it would be unwise to submit *Le monde* for publication.

Descartes later wrote and published another work containing

the law or principle of inertia and called *Principia philosophiae,* or *Principles of Philosophy.* In the meanwhile, the law had been published by Pierre Gassendi, a French philosopher and scientist. Gassendi also made experiments to test the law. These included dropping weights on moving carriages and ships.

Descartes's *Principia* had a tremendous influence; for example, it was a seminal work for Isaac Newton. The latter's *Principia* was named to make it appear as an improvement on Descartes's. Descartes had written a *Principia philosophiae,* but Newton went a step further in creating a *Philosophiae naturalis principia mathematica.* That is, Newton was not so much concerned with principles of philosophy in general as with natural philosophy or physical science, and its mathematical principles. In his formulation of the law of inertia, Newton even used certain expressions found in Descartes's *Principia,* such as *status* ("state") and *quantum in se est* ("as much as in it lies"). It would even seem that Newton's presentation of this law as the first of his "axioms, or laws of motion" *(axiomata, sive leges motus)* had been conditioned by Descartes's naming of his law as one of "certain rules, or laws of nature" *(regulae quaedam sive leges naturae).*

SUPPLEMENT
9

Galileo's Discovery of the Parabolic Path

Galileo's discovery of the parabolic path appears to have two parts. One is the mathematical proof that a projectile moving in a space free of resistance will have two independent components: a vertical component that follows the law of free fall (just as if there were no horizontal component) and a horizontal component of forward motion that is uniform (just as if there were no vertical component). Vertically, the distance fallen D_y is proportional to the square of the time T; horizontally, the distance D_x, through which the body advances, is proportional to the time. The combination of $D_y \propto T^2$ and $D_x \propto T$ yields a parabola (see Supplement 10, Sec. 8). Galileo knew the law of free fall ($D_y \propto T^2$) as early as 1604. Stillman Drake finds it to be "certain that Galileo discovered the parabolic trajectory no later than 1608 and proved it mathematically early in 1609." But Galileo did not mention this discovery in print until some thirty years later in the *Two New Sciences.*

Drake's reconstruction of Galileo's discovery is based on the interpretation of some diagrams and numerical data and calculations on some odd pages of Galileo's notes, without explanatory text. Drake shows that these notes are consistent with an experiment in which a ball rolls down an inclined plane and is then deflected so as to move out horizontally into space. Galileo, it is assumed, was testing horizontal inertial motion, and this device enabled him to fire balls out in a horizontal direction with a determinate speed. Drake concludes his analysis of these documents with the suggestion that Galileo must have observed the parabolic path as a by-product of these experiments. Drake's evidence was not merely that he could reproduce Galileo's numbers as calculated results, but that he was able to devise and construct an experimental set-up in which the data he recorded were "sufficiently similar to the data recorded by Galileo to verify the hypothesis that he experimentally obtained sets of numbers measured to three or four significant figures." If this is a correct analysis, we can only wonder that in his final discussion (in print) of parabolic trajectories, Galileo did not refer to any quantitative experiments or even hint that such were possible.

SUPPLEMENT
10

A Summary of Galileo's Major Discoveries in the Science of Motion

Galileo's *Two New Sciences* presents a mathematical theory of freely falling bodies, much of which he appears to have discovered some thirty years earlier. Galileo's major discoveries include the following:

1. Contrary to popular belief, a heavy and a light object do not fall from a high place (e.g., a tower) with speeds proportional to their weights, but with almost identical speeds.

2. If a body falls in air (or any other resisting medium), the resistance will increase as some function of the speed; when the resistance becomes equal to the body's weight, the acceleration will cease and the body will continue to move with uniform speed downward.

3. In limited circumstances (e.g., on a smooth horizontal plane or when the air's resistance equals and cancels the accelerating force of weight), a body will continue in a motion it has been given or has acquired. (Galileo supposed that this limited or restricted principle of inertia also applied to a large spherical surface concentric with the earth, e.g., the earth's surface. He also linked this principle with a body's tendency to maintain rotation.)

4. In natural acceleration, or in uniformly accelerated motion, the speed increases as the integers 1,2,3, ... (We write this law algebraically as, starting from rest, $V \propto T$ [or $V=AT$]). It follows that the distance increases as the square of the time, or $D \propto T^2$ (actually $D=1/2AT^2$). Galileo showed by experiment that $D \propto T^2$ is valid for the motion of a ball rolling down any inclined plane.

   (a) In such motion, the distances traversed in successive equal intervals of time are as the odd numbers 1,3,5,7, . . . because the total distances traversed are as the squares (1,4,9,16, . . .) and $4-1=3$, $9-4=5$, $16-9=7$, ...

5. Free fall and rolling down an inclined plane are examples of uniformly accelerated motion. Hence the laws of free fall are $V \propto T$ and $D \propto T^2$.

   (a) Falling in air is not an example of pure uniform acceleration, because of the resistance of air; this is the reason that when two bodies of unequal weight are dropped from a tower, the heavier one will strike the ground just a moment before the lighter one.

6. In motion along an inclined plane, the final velocity will be the same for all angles of inclination so long as the starting point is at the same height above the level.

   (a) The time of descent is the same along all chords of a vertical circle that end at the lowest point of the circle.

   (b) If a body is accelerated uniformly during a given time interval and then deflected so as to move uniformly with the speed so acquired, it will—during another such time interval—move uniformly through double the distance it moved under the initial acceleration.

7. Vertical and horizontal components of a compound motion are independent; hence a body (e.g., a projectile) can have a uniform horizontal component of motion and a uniformly accelerated vertical component, one being independent of the other.

8. The path of a projectile (neglecting the factor of air resistance) is a parabola. The reason is that the forward horizontal motion is uniform and the vertical motion is uniformly accelerated. In rectangular coordinates $x = V_0 T$ and $y = \frac{1}{2} AT^2$. Since V_0 and $\frac{1}{2} A$ are constants, say c and k, these equations become $x = cT$ and $y = kT^2$, whence $\frac{x^2}{c^2} = T^2$ and $\frac{y}{k} = T^2$ so that $y = Kx^2$ (where $K = \frac{k}{c^2}$), the equation of a parabola.

9. Galileo said that motion can be a "state" akin to a "state of rest," which is the equivalent of saying that a motion can continue indefinitely of and by itself, without the mediation of any external force. The concept of "state of motion" was developed by Descartes and became a cornerstone of Newton's edifice of rational mechanics.

We can see the truth of Galileo's "double distance rule" (6*b*) by using simple algebra. In uniformly accelerated motion during time T,

$$V = AT$$

$$D = \tfrac{1}{2} AT^2.$$

At the end of time T, let the body begin to move uniformly with the acquired speed V during another time-interval equal to T. The distance through which it will move is

$$\text{"distance"} = VT.$$

But since

$$V = AT$$

it follows that

$$\text{"distance"} = (AT) \times T = AT^2.$$

The quantity AT^2 is twice $\frac{1}{2} AT^2 = 2 (\frac{1}{2} AT^2) = 2D$.

In one of Galileo's early manuscripts, as interpreted by Stillman Drake, Galileo attempted to apply this correct result to the case of a ball rolling down an inclined plane and then deflected horizontally by a curved deflector. The observed distances did not then agree with the calculated distances. The reason is clear to us. There would have been agreement if there were frictionless sliding down the inclined plane, as would be the case for a sliding block of dry ice which floats on a cushion of carbon dioxide gas, or a piece of ordinary ice sliding down a very hot inclined plane. But Galileo had apparently been experimenting with a ball rolling down an inclined plane; he did not know that there would be a large discrepancy arising from the fact that the ball's motion and energy are not translational but include rotation (a factor of two-sevenths of the motion).

SUPPLEMENT
11

Newton's Debt to Hooke: The Analysis of Curvilinear Orbital Motion

In an acrimonious debate, in which Hooke wanted Newton to give him credit for the inverse-square law of gravity, Newton glossed over Hooke's real contribution to his thought. Hooke's contribution was not that he suggested the law of the inverse square, which Newton correctly believed followed simply enough (at least for circular orbits) from the analysis of circular motion, once the v^2/r law was known. What Hooke taught Newton was much more fundamental, namely, the correct way to analyze curvilinear motion.

In 1679, Hooke (recently appointed secretary of the Royal Society of London) wrote a friendly letter to Newton, expressing his hope that Newton would send some scientific communications to the society. Hooke then asked Newton to comment on what Hooke called a "hypothesis . . . of mine . . . of compounding the celestiall motions of the planetts [out] of a direct motion by the tangent & an attractive motion towards the centrall body." In his reply, Newton introduced another topic, but did not discuss Hooke's "hypothesis." In a later letter (6 Jan. 1680), Hooke wrote of "my supposition" concerning the force of attraction that keeps planets in their orbits: this "Attraction always is in a duplicate proportion to the Distance from the Center Reciprocall, and Consequently . . . the Velocity will be . . . as Kepler Supposes Reciprocall to the Distance." Speaking of the attraction as in "a duplicate proportion to the Distance . . . Reciprocall" is an old-fashioned way of saying the attraction is inversely proportional to the square of the distance. The velocity is here supposed to be inversely proportional to the distance. Hooke stressed to Newton that it was important to solve the problems of planetary motion and the motion of the moon, since such knowledge could lead to the solving of the problem of finding longitude at sea, which "will be of great Concerne to Mankind." Hooke was so proud of his letter to Newton that he read it aloud at a meeting of the Royal Society. Hooke reiterated his "supposition" in a later letter (17 Jan.) about "a centrall attractive power" and suggested that Newton's "excellent method" would "easily" enable him to "find out what that Curve must be" that results from this force and to "suggest a physicall Reason of this proportion."

In one of Newton's replies to Hooke, he stated plainly and simply that he had not ever heard of Hooke's "hypothesis" of compounding orbital motions out of a tangential motion and "an attractive motion towards the centrall body." We know that Newton himself had believed in a kind of centrifugal force, that is, a force associated with what seems a tendency of all bodies moving on curves to push or be pushed outward, away from the center.

Hooke's analysis contains the key to the study of celestial motions, and it became central to the development of Newton's celestial mechanics in the *Principia*. In many documents, Newton admitted that what started him off on this subject was his correspondence with Hooke. Newton gave a name to the centrally directed force: "centripetal." We have used it ever since. The kind of analysis that Newton made, using what he learned from Hooke, is shown in Fig. 32 on p. 168 and in Fig. 31 on p. 162.

Newton apparently worked up his first essay on celestial mechanics after the correspondence with Hooke in 1679/80. It was to this that he referred when he told Halley, during the latter's visit in 1684, that he had calculated the orbit of a planet under the action of an inverse-square force. But Newton did not need Hooke to tell him that the force varies as the inverse square of the distance. This follows from the simplest algebra (see pp. 165–66) once it is known that the force in circular motion is proportional to v^2/r; at least this is so for circular orbits and it would not be much of a guess that it is also so for ellipses. But, as Newton quite correctly assumed, it is one thing to make a good

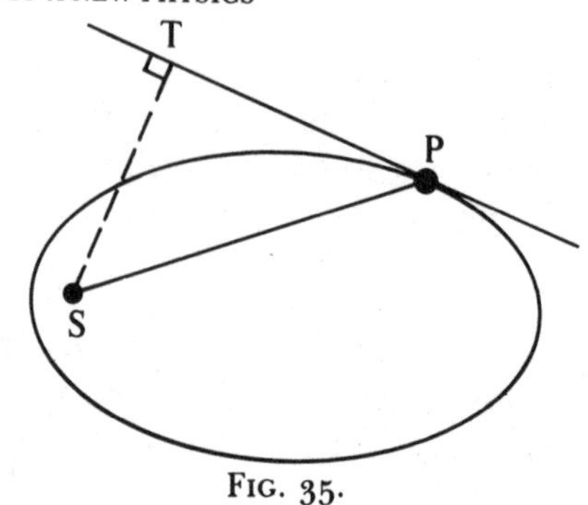

FIG. 35.

guess, and another to find out a mathematical truth and its consequences. It is easy to do the former and difficult to do the latter. He himself had guessed that the force would be as the inverse square, although he had been unfruitfully considering a centrifugal rather than a centripetal force. But he had known the v^2/r law long before Huygens published it in 1673.

Newton was fully aware that Hooke did not completely understand what he was writing about. Despite Hooke's keen analysis of curvilinear motion, he made an important error in concluding that the speed would be inversely proportional to the distance. As Newton readily proved, the speed is inversely proportional to the perpendicular to the tangent. In the diagram a planet is at *P*. What Hooke says is equivalent to the assertion that the speed at *P* is inversely proportional to the distance from the sun, *SP*, or

$$v \propto \frac{1}{SP},$$

but Newton says that the speed is rather inversely proportional to *ST*, the line drawn perpendicularly from the sun *S* to a point *T* on the tangent to the orbit at *P*,

$$v \propto \frac{1}{ST}.$$

Only at the apsides is Hooke's law true. Furthermore, Hooke's speed law is inconsistent with Kepler's law of areas. Kepler himself later found this out, whereupon he abandoned the inverse-distance speed law, which Hooke still believed was a true law for planetary orbital motion.

Newton was therefore correct in his judgment that Hooke did not really understand the consequences of his guess that the attractive force varies as the inverse square of the distance and that he did not therefore deserve credit for the law of Universal gravity. This would have seemed all the more true in that Newton was aware that he did not need Hooke to suggest to him the inverse-square character of the force. Hooke's claim to the inverse-square law has masked Newton's far more fundamental debt to him, the analysis of curvilinear orbital motion. In asking for too much credit, Hooke effectively denied to himself the credit due him for a seminal idea. (For further information see my *The Newtonian Revolution* [Cambridge and New York: Cambridge University Press, 1980, 1983], secs. 5.4, 5.5).

SUPPLEMENT 12

The Inertia of Planets and Comets

Newton's statement that the motion of planets and of comets illustrates the principle of inertia may seem puzzling, since their motion is curved. Newton expected his readers to understand that such motion has two components: a linear inertial motion along the tangent to the curve, and a continual accelerated motion "of falling" toward the center (centripetal) that keeps the motion along the curve rather than flying off on a tangent. Since the motion of planets and comets has continued for a very long time (undiminished by friction), and is likely to continue for a long time, the tangential component of their orbital motion provides the best example of inertial motion that continues on and on without sensible diminution. Terrestrial motions, such as those of projectiles, are not good examples because such motions are slowed down by air friction and do not last very long since all projectiles eventually fall down to the ground.

Newton also illustrated inertial motion by the spinning of a top or the rotation of the earth. In both cases, the individual particles of the rotating body have a linear tangential component of inertial motion, but because of the force of cohesion that holds the particles together, they do not fly off on a tangent. In fact, we know from experience how correct this analysis is, since many bodies can be made to spin so fast that they do fly apart; the reason is that their component parts are given so great a tangential velocity that the force of cohesion is no longer strong enough to keep them moving in a circular path. The situation would be similar if the moon were suddenly given a great increase in velocity. Then the force required to make the moon fall rapidly enough to stay in its orbit would be increased (according to the law of v^2/r). This force would become greater than the force of gravity that the earth exerts in its pull on the moon, and the moon would begin to fly off on a tangent.

SUPPLEMENT
13

Proof that an Elliptical Planetary Orbit Follows from the Inverse-Square Law

In a series of propositions (props. 11-13) in Book One of the *Principia,* Newton proves that if a planet moves in orbit in an ellipse, in a parabola, or in a hyperbola, the force varies inversely as the square of the distance from a focus. To do so, he invokes the law of areas (props. 1,2,3) and a very original mathematical measure of a force (prop. 6). Then, in the first edition, in corollary 1 to props. 11–13, Newton states, but does not prove, the converse of props. 11–13: given an inverse-square force, the orbit will be a conic section. In the subsequent prop. 17, Newton shows which condition yields a circle, an ellipse, a parabola, or a hyperbola, when the force varies as the inverse square of the distance. In the second edition of the *Principia,* Newton added the steps of the proof of the corollary to props. 11–13.

Many writers have confused the two propositions: (a) that a conic section implies an inverse-square force and (b) that an inverse-square force implies a conic section. The proof of one does not in and of itself imply the proof of the other. Newton himself was fully aware that a proof that "A implies B" does not prove that "B implies A." For example, in prop. 1 of the *Principia,* he proves that if there is a centripetal force acting on a body with an initial component of inertial motion, the law of areas holds true; but he then introduces prop. 2, to prove the converse, that the law of areas implies a centripetal force. In the first edition of the *Principia,* Newton did not actually give his proof that an inverse-square force implies an elliptical planetary orbit, but that does not necessarily mean either that he did not think such a proof was needed or that he had no such proof in his mind. The *Principia* is a very idiosyncratic book. Much of what Newton omitted as "obvious" is far from obvious to his readers, and yet there are times when he seems to go on at great length about what seems to us to be obvious or trivial.

What Newton seems to have proved after his correspondence with Hooke (see Supplement 11) is that an elliptical orbit implies an inverse-square law, and the tract that he wrote out after Halley's visit in 1684 proves this proposition. This is the case also in the first edition of the *Principia.* And yet, according to Conduitt's story of Halley's visit, Halley asked Newton what the orbit of a planet would be, given an inverse-square force (not what the force would be, given an elliptical orbit) and Newton replied that the path would be an ellipse and that he had "calculated it." Of course, this is Conduitt's recollection of what Halley had told him about a conversation with Newton many years earlier. We cannot be sure that it is an accurate record of what either Halley or Newton said on that famous occasion. In a later attempt to set forth the record of his development, Newton did say that in 1676–77 (an error for 1679–80) he "found the Proposition that by a centrifugal force [read centripetal force] reciprocally as the square of the distance a planet must Revolve in an Ellipsis about the center of the force placed in the lower umbilicus [or focus] of the Ellipsis or with a radius drawn to that center describe areas proportional to the times."

Several conclusions are possible, among them: (1) Newton proved that the ellipse implies an inverse-square force and mistakenly thought he had also proved the converse; (2) Newton proved that the ellipse implies an inverse-square force and worked out (in his mind or on paper) the proof of the converse; (3) Newton did not understand what he had proved and thought he had proved that an inverse-square force implies an elliptical planetary orbit; (4) Newton proved that the ellipse implies an inverse-square force and merely took it for granted that it was possible to prove the converse. It is not profitable to hypothesize about possible history in the absence of all evidence. But it is not very likely, in my opinion and that of other Newtonian scholars, that Newton would have committed the blunder of (1), an obvious logical fallacy. Similarly, it is unthinkable that a mathematician of Newton's ability would have made the error of (3). But (2) and (4) are possible. We do know that Newton was criticized for not having given a proof, in the first edition of the *Principia,* that an inverse-square law implies an elliptical orbit; he therefore revised the first corollary to props. 11–13 in the second edition, providing a proof.* On at least one occasion, Newton himself discussed this question. In an unpublished history of the development of the *Principia,* Newton wrote: "The Demonstration of the first Corollary of the 11th 12th & 13th Propositions being very obvious, I omitted it in the first edition . . ."

The facts, then, are that in print in the first edition, Newton stated (but gave no proof) that the inverse-square law implies an elliptical orbit; in the second edition, he supplied a proof. We can only guess or make hypotheses concerning this sequence. As Newton said, we should not build knowledge upon hypotheses.

*Scholars are grateful to Robert Weinstock (*American Journal of Physics* 50, pp. 610–17) for having brought this problem to their attention. But there is not agreement with his extreme position, corresponding to possibility (1). It is a wholly separate question whether the proof that Newton gives is or is not rigorous or even sound; Professor Weinstock holds that it is not a real proof at all.

Newton's autobiographical statements are collected (and transcribed) in Appendix I to my *Introduction to Newton's 'Principia'* (Cambridge, Mass.: Harvard University Press, 1971). They were written, during Newton's quarrels over priority, in order to establish what we know from other evidence to be an incorrect chronology of discovery, and so must be taken with a grain of salt. On this topic, see my *The Newtonian Revolution* (Cambridge and New York: Cambridge University Press, 1980, 1983), 248–49.

SUPPLEMENT
14

Newton and the Apple: Newton's Discovery of the v^2/r Law

Newton devoted much time and energy to composing and advancing a chronology of his discoveries that would place many of them at an earlier date than the primary historical documents would warrant. The reason for his imposition upon history of an imagined chronology was perhaps to date his discoveries so early that he could successfully combat his opponents in the controversies that arose over priority.

Newton may have invented the story of the apple, which would be dated in the mid-1660s, when he alleged he had made the moon test. We know that he himself told the story of the apple's falling, the origin of the oft-repeated statement that this was the occasion for his thinking about gravity's extending to the moon. He may also have come to believe, later in life and long after the event, that he had calculated the falling of the moon in the 1660s and had found the test to agree approximately. But what he was actually calculating was not the falling of the moon, as in the famous moon test given in the scholium to prop. 4 of Book Three of the *Principia,* but something quite different.*

With regard to Newton's early discovery of the v^2/r law for uniform circular motion, we are on better grounds. At this time Newton was seeking a measure of "centrifugal endeavour"; only

*On Newton's calculations and their significance, see my monograph on "The *Principia,* Universal Gravitation, and the 'Newtonian Style' " in Zev Bechler, ed., *Contemporary Newtonian Research* (Dordrecht [Holland] and Boston: D. Reidel Publishing Co., 1982); and sec. 5.3 of my *The Newtonian Revolution.*

later, in 1680 (see Supplement 11), did Newton become converted by Hooke to the concept of centripetal force. After Halley informed Newton that Hooke wanted to be given credit for the inverse-square law, Newton sent Halley an outline of a demonstration, based on his research of some twenty years earlier, to be added at the end of the scholium following prop. 4 of Book One of the *Principia.* Newton wanted it to be evident to all readers that he knew the v^2/r law earlier (in fact, almost a decade earlier) than the publication of the law by Christiaan Huygens in his *Horologium oscillatorium* of 1673. Since prop. 4 deals with uniform circular motion, Newton was saying in effect that he long ago had learned that the force in this case is as v^2/r and hence it would be easy to show (by a little algebra plus Kepler's third law) that the force is as $1/r^2$. So he would not have needed Hooke to tell him twenty years later about a $1/r^2$ force.

In 1960 John Herivel analyzed some early writings of Newton's on force and motion, entered by Newton in a "Waste Book" in early 1665 or soon thereafter. Herivel showed that in this document Newton derived the v^2/r law in a very original way.* Hence there can be no doubt that Newton had found this law very early and completely independently of Huygens.

*See John W. Herivel's "Newton's Discovery of the Law of Centrifugal Force," *Isis* 51 (1960), 546–53; also Herivel's *The Background to Newton's Principia* (Oxford: Clarendon Press, 1965), 7–13.

SUPPLEMENT
15

Newton on "Gravitational" and "Inertial" Mass

In the derivation on pp. 169–71, two equations are given for the force acting on a terrestrial object such as an apple. One is the *gravitational* equation,

$$F = G\frac{mM_e}{R_e^2}$$

or

$$W = G\frac{mM_e}{R_e^2}$$

and the other is the *dynamical* or *inertial* equation,

$$m = \frac{F}{A}$$

or

$$F = mA$$

which, in the case of the weight force, became

$$W = mA.$$

It should be observed that in the second group of equations, the quantity m is a measure of the body's inertia, that is, the *inertial* resistance (F/A) to the body's being accelerated or undergoing a change in its state of motion or of rest. To be precise, let us give this quantity a special twentieth-century name, "inertial mass," and replace the symbol m by m_i to denote its inertial quality. The final equations above should now be rewritten as

$$F = m_i A$$

$$W = m_i A.$$

Let us now consider the quantity m (or the mass) that appears in the first group of equations

$$F = G\frac{mM_e}{R_e^2}.$$

Here m has no obvious connection with the *inertial* resistance of the body to being accelerated, to undergoing a change in state. Rather, this quantity is a measure (or is the determining factor) of the body's *gravitational* response to the earth. Or, to use the language of our present physics, this is a measure (or the determinant) of the body's response to the earth's gravitational field (or to any other gravitational field). It thus may be given the twentieth-century name "gravitational mass." Accordingly, we may use the symbol m_g for this quantity. The first two equations now become

$$F = G\frac{m_g M_e}{R_e^2}$$

$$W = G\frac{m_g M_e}{R_e^2}.$$

When we equate the two expressions for W, we now get

$$m_i A = G\frac{m_g M_e}{R_e^2}.$$

To cancel out the factor of m is to suppose that

$$m_i = m_g$$

a step that requires further analysis. Does m_i always equal m_g?

Both varieties of mass—*gravitational* mass and *inertial* mass—correspond to our intuitive concept of "quantity of matter." In the case of a pure substance such as aluminum, both would be proportional to the volume of aluminum (which would be a measure of the amount or "quantity" of aluminum). The conceptual problem can be stated as follows: Is there any logical reason why a body's response to a gravitational field (or its gravitational mass) should be the same as its resistance to being accelerated by non-gravitational as well as by gravitational forces (or its inertial mass)? In fact, in the framework of Newtonian or "classical" physics, the answer is a flat No! It is only in post-Newtonian or relativistic physics that there is a necessary "equivalence" of gravitational and inertial mass. How then did Newton deal with this problem?

Before presenting Newton's solution, let us observe the very high level of consideration to which Newton's physics leads. Whereas Galileo was concerned with a body's weight, Newton has introduced a very different and modern concept of mass. This concept is original with Newton, although there are (as in any novel scientific concept) some antecedents to be found—for instance, in Kepler's writings about *moles* (a kind of "bulk") and in certain discussions by Huygens.

If the equivalence of inertial and gravitational mass does not follow from logic, and is not an integral part of theory, then the only way in which it can be known is from experiment. Newton first recognized the need for making such an experiment some time in 1685, after completing the first version of his tract *De motu.* The experiment would use two equal pendulums, with bobs containing varieties of matter; any variation in the ratio of inertial to gravitational mass would show up as a difference in the period of vibration. Shortly afterwards, in a set of *De motu corporum definitiones (Definitions Concerning the Motion of Bodies),* he lists the substances on which he has made the experiment: gold, silver, lead, glass, sand, common salt, water, wood, and wheat. In both the preliminary and final versions of Book Three of the *Principia* (prop. 6 of Book Three, sec. 9 of *System of the World*), Newton describes the experiment in detail. He has constructed two long pendulums of identical length, containing hollow boxlike bobs, at the center of which he can place equal amounts of these nine substances. Since the pendulums have identical bobs, both encounter the same factor of air resistance. He has shown mathematically that the existence of the same period of vibration for bobs containing equal amounts of these nine substances proves that their weights are proportional to their quantities of matter. By simple induction Newton proceeds to the general law.

Newton describes matter in terms of its *weight* and its *quantity;* the latter is, for Newton, the inertia of matter. The expressions *gravitational* and *inertial* mass were introduced into physics in Albert Einstein's writings on relativity. Newton, furthermore, did not write out equations in the way that I have done. But he did essentially proceed to develop this topic in the way that these equations symbolize. Thus he concluded that his pendulum experiments had shown with very great accuracy the results long observed (going back to Galileo's "tower" experiments) that, but for the small factor of air resistance, all sorts of heavy bodies fall to the earth from equal heights in equal times.

One of the reasons why Newton's concept of mass is so important is that mass is a fundamental or permanent property of bodies, whereas weight is an accidental property. Newton's physics showed, for example, that a body's weight (or the effect on a body of the earth's pull) could vary with its position on the earth, the weight being a calculable property of geographical latitude. A body, additionally, would weigh less out in space than on the earth's surface, according to the inverse-square law. Furthermore, a body's "weight" toward the moon on the moon's surface would be notably different from what it would be on the earth's surface. But wherever in space a body might happen to be, its mass (according to Newtonian physics) is ever the same—both its inertial mass (its resistance to being accelerated) and its gravitational mass. Furthermore, mass is an important property to consider in relation to bodies out in space—sun, planets, moons, and stars—even though their "weight" (in the sense of the earth's pull on them) is not of consequence. By shifting the discussions of physics from weight to mass, Newton made possible a universal science in place of a local terrestrial science.

Most readers are aware that in relativistic physics, mass is no longer conceived to be an independent constant of a body. Rather, it turns out to be related to the body's speed relative to the frame of reference. But for ordinary bodies (that is, those whose speeds are small with respect to the speed of light), the difference between the two is negligible.

SUPPLEMENT
16

Newton's Steps to Universal Gravity

In the autumn of 1684, following Halley's visit, Newton composed his tract *De motu,* in which he proved the following proposition: Motion in an ellipse according to the law of areas requires that a central or centripetal inverse-square force be directed toward the point with respect to which the equal areas are reckoned. Since the planets move in elliptical orbits, with the sun at a focus, and since a line from the sun to a planet sweeps out equal areas in any equal times, Newton concluded (1) that there must be a force directed from each planet to the sun, and (2) that this sun-directed force varies inversely as the square of the distance. In evident pride on discovering the planetary law of force, he wrote that he had proved that, to quote his own words, "the major planets revolve in ellipses having a focus in the center of the sun, and radii drawn [from the planets] to the sun describe areas proportional to the times, entirely as Kepler supposed . . . ."

In fact, Newton had not proved this proposition, nor did he continue to believe it for long. Strictly speaking, it is false. As Newton soon realized, the planets do not move exactly according to the law of areas in simple Keplerian elliptical orbits with the sun at a focus. Instead, the focus lies in the common center of mass, because not only does the sun attract each planet but also each planet attracts the sun (and the planets attract one another). If Newton had already formulated his principle of universal gravitation, he would not have proposed, and have assumed he had proved, this erroneous proposition.

Newton very quickly became aware that he had not proved that the planets move precisely according to the law of elliptical orbits and the law of areas. He had only found that these two planetary laws of Kepler hold for a one-body "system": a single point mass moving with an initial component of inertial motion in a central-force field. He recognized that the one-body system corresponds not to the real world but to an artificial situation that is easier to investigate mathematically. The one-body system reduces the earth to a point mass and the sun to an immobile center of force.

In his premature jubilation, Newton had neglected to take into consideration what we know today as the Newtonian third law of motion: that to every action there must be an equal and opposite reaction. In other words, if a body A exerts a force on body B, then body B must simultaneously exert an equal and opposite force on body A. In the case of the sun and a planet, say the earth, this law implies that if the sun exerts a force on the earth to keep the earth in orbit, then the earth must exert an equal force on the sun. In theory, each of these two bodies pulls on the other, with the result that each must move in an orbit about their common center of gravity. Since the mass of the earth is so minuscule compared to the sun's mass, their common center of gravity is practically at the center of the sun and the sun's motion is virtually non-existent. But this is not the case for the sun and Jupiter, the most massive planet in the solar system, nor for the earth and the moon.

The development of Newton's thinking on action and reaction after he completed the first draft of *De Motu* is set out in the opening sections of the first book of the *Principia.* In the introduction to the 11th section Newton explains that he has confined himself so far to a situation that "hardly exists in the real world," namely the "motions of bodies attracted toward an unmoving center." The situation is artificial because "attractions customarily are directed toward bodies and—by the third law of motion—the actions of attracting and attracted bodies are always mutual and equal." As a result, "if there are two bodies, neither the attracting nor the attracted body can be at rest." Rather, "both bodies (by the fourth corollary of the laws) revolve about a common center, as if by a mutual attraction."

Newton had seen that if the sun pulls on the earth, the earth must also pull on the sun with a force of equal magnitude. In this two-body system the earth does not move in a simple orbit around the sun. Instead the sun and the earth each move about their mutual center of gravity. A further consequence of the third law of motion is that each planet is a center of attractive force as well as an attracted body; it follows that a planet not only attracts and is attracted by the sun but also attracts and is attracted by each of the other planets. Here Newton has taken the momentous step from an interactive two-body system to an interactive many-body system.

In December 1684, Newton completed a revised draft of *De motu* that describes planetary motion in the context of an interactive many-body system. Unlike the earlier draft, the revised one concludes that "the planets neither move exactly in ellipses nor revolve twice in the same orbit." This conclusion led Newton to the following result: "There are as many orbits to a planet as it has revolutions, as in the motion of the Moon, and the orbit of any one planet depends on the combined motion of all the planets, not to mention the actions of all these on each other." He then wrote: "To consider simultaneously all these causes of motion and to define these motions by exact laws allowing of convenient calculation exceeds, unless I am mistaken, the force of the entire human intellect."

There are no documents that indicate how, in the month or so between writing the first draft of *De motu* and revising it, Newton came to perceive that the planets act gravitationally on one another. Nevertheless, the passage cited above expresses this perception in unambiguous language: *eorum omnium actiones in se invicem* ("the actions of all these on each other"). A consequence of this mutual gravitational attraction is that all three of Kepler's laws are not strictly true in the world of physics but are true only for a mathematical construct in which point masses that do not interact with one another orbit either a mathematical center of force or a stationary attracting body. The distinction Newton draws between the realm of mathematics, in which Kepler's laws are truly laws, and the realm of physics, in which they are only "hypotheses," or approximations, is one of the revolutionary features of Newtonian celestial dynamics.

In the spring of 1685, a few months after revising *De motu*, Newton was well on his way to finishing the first draft of the *Principia.* In the initial version of what was to become a second book, "The System of the World," he spelled out the steps that led him to the concept of planetary gravitational interactions. In these steps the third law of motion has the chief role. There is no reason to believe they are not the same steps that led him to the same concept a few months earlier when he revised *De motu.*

Here are two passages from the first draft of the *System of the World* (newly translated from the Latin by Anne Whitman and me) that bring out the crucial role of the third law of motion:

> 20. *The agreement between the analogies.*
> And since the action of centripetal force upon the attracted body, at equal distances, is proportional to the matter in this body, it is reasonable, too, that it is also proportional to the matter in the attracting body. For the action is mutual, and causes the bodies by a mutual endeavor (by law 3) to approach each other, and accordingly it ought to be similar to itself in both bodies. One body can be considered as attracting and the other as attracted, but this distinction is more mathematical than natural. The attraction is really that of either of the two bodies toward the other, and thus is of the same kind in each of the bodies.
>
> 21. *And their coincidence.*
> And hence it is that the attractive force is found in both bodies. The sun attracts Jupiter and the other planets, Jupiter attracts its satellites and similarly the satellites act on one another and on Jupiter, and all the planets on one another. And although the actions of each of a pair of planets on the other can be distinguished from each other and can be considered as two actions by which each attracts the other, yet inasmuch as they are between the same two bodies they are not two but a simple operation between two termini. Two bodies can be drawn to each other by the contraction of one rope between them. The cause of the action is twofold, namely, the disposition of each of the two bodies; the action is likewise twofold, insofar as it is upon two bodies; but insofar as it is between two bodies it is single and one. There is not, for example, one operation by which the sun attracts Jupiter and another operation by which Jupiter attracts the sun, but one operation by which the sun and Jupiter endeavor to approach each other. By the action by which the sun attracts Jupiter, Jupiter and the sun endeavor to approach each other (by law 3), and by the action by which Jupiter attracts the sun, Jupiter and the sun also endeavor to approach each

other. Moreover, the sun is not attracted by a twofold action toward Jupiter, nor is Jupiter attracted by a twofold action toward the sun, but there is one action between them by which both approach each other.

Next Newton concluded that "according to this law all bodies must attract each other." He proudly presented the conclusion and explained why the magnitude of the attractive force is so small that it is unobservable. "It is possible," he wrote, "to observe these forces only in the huge bodies of the planets."

In Book Three of the *Principia,* which is also concerned with the system of the world but is somewhat more mathematical, Newton treats the topic of gravitation in essentially the same way. First, in what is called the moon test, he extends the weight force, or terrestrial gravity, to the moon and demonstrates that the force varies inversely with the square of the distance. Then he identifies the same terrestrial force with the force of the sun on the planets and the force of a planet on its satellites. All these forces he now calls gravity. With the aid of the third law of motion he transforms the concept of a solar force on the planets into the concept of a mutual force between the sun and the planets. Similarly, he transforms the concept of a planetary force on the satellites into the concept of a mutual force between planets and their satellites and between satellites. The final intellectual transformation yields the universal principle that all bodies interact gravitationally.

We can see how Newton's creative imagination directed him toward the concept of universal gravity. The same argument that led him to interplanetary forces can be applied to satellite systems, the earth, and an apple. Since all apples must be bodies which both originate a gravitational pull and react to a gravitational pull, they must pull on one another. Eventually this train of thought leads to the bold conclusion that any two bodies anywhere in the universe act gravitationally on each other. Thus the logic of physics, guided by a creative mathematical intuition, produces a law of mutual force that applies to all bodies, terrestrial or celestial, wherever they may happen to be. This force varies inversely as the square of the distance and is proportional to the gravitating masses:

$$F \propto \frac{m_1 m_2}{D^2}$$

or

$$F = G\frac{m_1 m_2}{D^2}$$

where m_1 and m_2 are the masses, D the distance between them, and G the universal constant of gravitation.

This analysis of the stages of Newton's thinking should not be taken as in any way diminishing the extraordinary force of his creative genius; rather, it should make that genius plausible. The analysis shows Newton's fecund way of thinking about physics, in which mathematics is applied to the external world as it is revealed by experiment and critical observation. This mode of creative scientific reasoning, which has been called "the Newtonian style," is captured by the English title of Newton's great work: *Mathematical Principles of Natural Philosophy.*

A Guide to Further Reading

Asterisks designate works from which quotations have been taken, with the publishers' permission, for inclusion in this book.

GENERAL BACKGROUND & EARLY SCIENCE

Marshall Clagett. *Greek Science in Antiquity.* New York: Abelard-Schuman, 1955. Revised reprint, New York: Collier Books; London: Collier-Macmillan, 1966.

O. Neugebauer. *The Exact Sciences in Antiquity.* Princeton: Princeton University Press, 1952; 2d ed., Providence, R.I.: Brown University Press, 1957; New York: Harper Torchbooks, 1962. Also *Astronomy and History: Selected Essays.* New York and Berlin: Springer-Verlag, 1983.

*Sir Thomas Little Heath. *Aristarchus of Samos, the Ancient Copernicus: A History of Greek Astronomy to Aristarchus.* Oxford: Clarendon Press, 1913.

Edward Grant. *Physical Science in the Middle Ages.* New York and London: John Wiley & Sons, 1971; Cambridge: Cambridge University Press, 1981.

Alistair C. Crombie. *Medieval and Early Modern Science.* 2 vols. 2d ed. Garden City, N.Y.: Doubleday Anchor Books, 1959. Also issued as *Augustine to Galileo*, 1952, 1961, 1979, etc.

THE SCIENTIFIC REVOLUTION

Marie Boas. *The Scientific Renaissance 1450–1630.* New York: Harper & Brothers, 1962; Harper Torchbooks, 1966.

Herbert Butterfield. *The Origins of Modern Science.* 2d ed. New York: Macmillan Co., 1957.

Richard S. Westfall. *The Construction of Modern Science: Mechanisms and*

Mechanics. New York and London: John Wiley & Sons, 1971; Cambridge: Cambridge University Press, 1978.

A. Rupert Hall. *The Scientific Revolution, 1500–1800: The Formation of the Modern Scientific Attitude.* London and New York: Longmans, Green and Co., 1954; Boston: Beacon Press, 1956. 2d ed., 1962. Revised ed., *The Revolution in Science, 1500–1750.* London and New York: Longmans, 1983.

I. Bernard Cohen. *Revolution in Science*. Cambridge, Mass., and London: Harvard University Press, 1985.

ASTRONOMY AND COSMOLOGY

J. L. E. Dreyer. *History of the Planetary Systems from Thales to Kepler.* Cambridge: Cambridge University Press, 1906. Reprint, under new title, *A History of Astronomy from Thales to Kepler.* New York: Dover Publications, 1953.

Alexandre Koyré. *From the Closed World to the Infinite Universe.* Baltimore: Johns Hopkins Press, 1957; New York: Harper Torchbooks, 1958.

Thomas S. Kuhn. *The Copernican Revolution: Planetary Astronomy in the Development of Western Thought.* Cambridge, Mass.: Harvard University Press, 1957.

THE WORK OF COPERNICUS

Edward Rosen. *Three Copernican Treatises.* New York: Columbia University Press, 1939. Contains translations of the *Commentariolus* of Copernicus, *Letter against Werner,* and Rheticus's *Narratio prima,* with commentaries and introduction. A third edition, revised, contains a biography of Copernicus plus Copernicus bibliographies 1939–1958 and 1959–1970. New York: Octagon Books, 1971.

Noel M. Swerdlow. "The Derivation and First Draft of Copernicus's Planetary Theory: A Translation of the *Commentariolus* with Commentary." *Proceedings of the American Philosophical Society* 117 (1973), 423–512.

Nicholas Copernicus. *On the Revolutions.* Ed. Jerzy Dobrzycki, translation and commentary by Edward Rosen. London: Macmillan; Baltimore: The Johns Hopkins University Press, 1978.

N. M. Swerdlow and O. Neugebauer. *Mathematical Astronomy in Copernicus's* De revolutionibus. 2 vols. New York and Berlin: Springer-Verlag, 1984.

Marjorie Nicolson. *Science and Imagination*. Ithaca: Cornell University Press, 1956; Hamden, Conn.: Archon Books, 1976. Deals with the effects of the telescope and of the "new astronomy" in general on the imagination and, especially, on English literature.

Ludovico Geymonat. *Galileo Galilei: A Biography and Inquiry into his Philosophy of Science.* Foreword by Giorgio de Santillana. Text translated from the Italian with additional notes and appendix by Stillman Drake. New York and London: McGraw-Hill Book Company, 1965.

Ernan McMullin, ed. *Galileo: Man of Science.* New York and London: Basic Books, 1967. Contains articles on various aspects of Galileo's life, work, and influence.

Stillman Drake. *Galileo at Work: His Scientific Biography.* Chicago and London: The University of Chicago Press, 1978.

*Stillman Drake, tr. and ed. *Discoveries and Opinions of Galileo.* Garden City, N.Y.: Doubleday Anchor Books, 1957. Contains translations of Galileo's *The Starry Messenger* (1610), *Letters on Sunspots* [i.e., *History and Demonstrations Concerning Sunspots and Their Phenomena*] (1613), *Letter to the Grand Duchess Christina* (1615), . . . with commentaries and introductions.

*Galileo Galilei. *Dialogue Concerning the Two Chief World Systems—Ptolemaic and Copernican.* Translated by Stillman Drake. Berkeley and Los Angeles: University of California Press, 1953. Revised reprint 1962. Another version, *Dialogue on the Great World Systems, in the Salusbury Translation,* 1661. Revised and annotated by Giorgio de Santillana. Chicago: The University of Chicago Press, 1953.

*Galileo Galilei. *Two New Sciences: Including Centers of Gravity & Force of Percussion.* Translated, with introduction and notes, by Stillman Drake. Madison: The University of Wisconsin Press, 1974. An earlier translation, by Henry Crew and Alfonso de Salvio, contains numerous errors and misleading interpretations.

Giorgio de Santillana. *The Crime of Galileo.* Chicago: The University of Chicago Press, 1955.

Jerome J. Langford. *Galileo, Science, and the Church.* Revised edition. Ann Arbor: The University of Michigan Press, 1971.

Winifred L. Wisan. "The New Science of Motion: A Study of Galileo's *De motu locali.*" *Archive for History of Exact Sciences* 13 (1974), 103–306. Also "Galileo and the Process of Scientific Creation." *Isis* 75 (1984),

269–86. Winifred Wisan, like R. Naylor (for whom see Supplement 4, p. 199), interprets Galileo's manuscripts in a way that differs radically from Stillman Drake's readings.

M. Segre. "The Role of Experiment in Galileo's Physics." *Archive for History of Exact Sciences* 23 (1980), 227–52. A critical summary of evidence and interpretations.

Albert Van Helden. *The Invention of the Telescope.* Philadelphia: The American Philosophical Society, 1977 (*Transactions of the American Philosophical Society,* vol. 67, pt. 4).

THE WORK OF KEPLER

Max Caspar. *Kepler.* Translated by C. Doris Hellman. New York and London: Abelard-Schuman, 1959.

Gerald Holton. "Johannes Kepler's Universe: Its Physics and Metaphysics." *American Journal of Physics* 24 (1956), 340–51.

Edward Rosen, translator and commentator. *Kepler's Somnium: The Dream, or Posthumous Work on Lunar Astronomy.* Madison, Milwaukee, and London: The University of Wisconsin Press, 1967. *Kepler's Conversation with Galileo's Sidereal Messenger,* with introduction and notes. New York and London: Johnson Reprint Corporation, 1965.

Johannes Kepler. *Mysterium cosmographicum: The Secret of the Universe.* Translation by A. M. Duncan, introduction and commentary by E. J. Aiton, with a preface by I. Bernard Cohen. New York: Abaris Books, 1981.

Arthur Koestler. *The Watershed: A Biography of Johannes Kepler.* Garden City, N.Y.: Doubleday & Company, Anchor Books, 1960.

Owen Gingerich. "Kepler, Johannes." *Dictionary of Scientific Biography.* Edited by Charles C. Gillispie. Vol. 7. New York: Charles Scribner's Sons, 1973, 289–312.

Arthur Beer and Peter Beer, eds. *Kepler: Four Hundred Years. Proceedings of Conferences Held in Honour of Johannes Kepler. Vistas in Astronomy,* 18. Oxford and New York: Pergamon Press, 1975. A mammoth work (1034 pages), containing extracts, summaries, and articles on every imaginable aspect of Kepler's life, work, and influence; contains three general summary articles: W. Gerlach's "Johannes Kepler—Life, Man and Work" (pp. 73–95), Martha List's "Kepler as a Man" (pp. 97–105), and I. B. Cohen's "Kepler's Century: Prelude to Newton's" (pp. 3–36).

THE LIFE AND WORK OF NEWTON

R. S. Westfall. *Never at Rest: A Biography of Isaac Newton.* Cambridge, London, and New York: Cambridge University Press, 1980.

Gale E. Christianson. *In the Presence of the Creator: Isaac Newton and His Times.* New York: The Free Press; London: Collier Macmillan Publishers, 1984).

I. Bernard Cohen. *Introduction to Newton's 'Principia'.* Cambridge, Mass.: Harvard University Press; Cambridge: Cambridge University Press, 1971.

I. Bernard Cohen. *The Newtonian Revolution, with Illustrations of the Transformation of Scientific Ideas.* Cambridge, London, and New York: Cambridge University Press, 1980.

Isaac Newton. *Mathematical Principles of Natural Philosophy.* Translated by Andrew Motte (1729). Revised by Florian Cajori. Berkeley: University of California Press, 1934. A new translation by I. Bernard Cohen and Anne Miller Whitman is scheduled for publication in 1986 by Harvard University Press and Cambridge University Press.

Isaac Newton. *Opticks, or a Treatise of the Reflections, Refractions, Inflections & Colours of Light* [1704; 4th ed. 1730]. Reprinted with foreword by Albert Einstein, introduction by Sir Edmund Whittaker, preface by I. Bernard Cohen, and analytical table of contents by Duane H. D. Roller. New York: Dover Publications, 1952; revised printing 1982. A scholarly edition of the *Opticks* by Henry Guerlac is scheduled for publication by Cornell University Press in 1985.

Isaac Newton's Papers & Letters on Natural Philosophy. Edited by I. Bernard Cohen and Robert E. Schofield. Cambridge, Mass.: Harvard University Press, 1958. Revised edition 1978.

ADDITIONAL SOURCES

*The *Paradiso* of Dante Alighieri. The Temple Classics. London: J. M. Dent, 1899, 1930.

*M. R. Cohen and I. E. Drabkin. *A Source Book in Greek Science.* Cambridge, Mass.: Harvard University Press, 1958.

*W. K. C. Guthrie, tr. Aristotle's *On the Heavens.* Loeb Classical Library. Cambridge, Mass.: Harvard University Press, 1939.

*E. W. Webster, tr. Aristotle's *Meteorologica.* Oxford: Clarendon Press, 1931. Also in *The Works of Aristotle Tanslated into English*, ed. W.D. Ross, Vol. 3.

*John F. Dobson and Selig Brodetsky, trs. "Preface and Book I" of Copernicus's *De revolutionibus. Occasional Notes of the Royal Astronomical Society* 2 (1947), 1–32.

*Johannes Kepler. *The Harmonies of the World,* Book 5, translated by Charles Glenn Wallis. Great Books of the Western World, 16. Chicago: Encyclopaedia Britannica, 1952.

**The Principal Works of Simon Stevin.* Vol. 1. *General Introduction, Mechanics.* Edited by E. J. Dijksterhuis. Amsterdam: C. V. Swets and Zeitlinger, 1955.

Index